Perl for Exploring DNA

Perl for Exploring DNA

Mark D. LeBlanc and Betsey Dexter Dyer

OXFORD

UNIVERSITY PRESS

2007

OXFORD
UNIVERSITY PRESS

Oxford University Press, Inc., publishes works that further
Oxford University's objective of excellence
in research, scholarship, and education.

Oxford New York
Auckland Cape Town Dar es Salaam Hong Kong Karachi
Kuala Lumpur Madrid Melbourne Mexico City Nairobi
New Delhi Shanghai Taipei Toronto

With offices in
Argentina Austria Brazil Chile Czech Republic France Greece
Guatemala Hungary Italy Japan Poland Portugal Singapore
South Korea Switzerland Thailand Turkey Ukraine Vietnam

Published by Oxford University Press, Inc.
198 Madison Avenue, New York, New York 10016

www.oup.com

Oxford is a registered trademark of Oxford University Press

Library of Congress Cataloging-in-Publication Data
LeBlanc, Mark (Mark D.), 1962–
Perl for exploring DNA/Mark LeBlanc and Betsey Dyer.
p. cm.
Includes bibliographical references.
ISBN 978-0-19-532757-1;   978-0-19-530589-0 (pbk.)
1. Nucleotide sequence—Data processing. 2. Amino acid sequence—Data processing.
3. Perl (Computer program language) I. Dyer, Betsey Dexter. II. Title
QP625.N89L43 2007
572.6'.633—dc22   2006036861

1 3 5 7 9 8 6 4 2

Printed in the United States of America
on acid-free paper

To my parents, Robert and Patricia;
my wife and best friend, Kathleen;
and sons, Jacob, Zachary, David, Jonathan,
Joshua, and Nathan

—Mark D. LeBlanc

To my patient, encouraging, and indulgent family:
Robert, Alice, and Samuel Obar

—Betsey Dexter Dyer

Acknowledgments

Gregory Williams was our Perl guru. A project like this should never be undertaken without one. Greg tried out all the code, added numerous creative suggestions, and made brilliant adjustments. If you notice an especially elegant section of Perl in this book, it probably came from Greg's bag of tricks, accumulated over years of using Perl to solve problems.

Daniel Saffioti, Mark's Australian mate and all around software wizard, was especially encouraging as he read through and commented on early drafts of the manuscript. One reviewer, Michael Raymer, was especially helpful in his detailed review. Many of his constructive suggestions are reflected in the text. Other reviewers who remained anonymous made significant improvements to early drafts.

Nathan Buggia and Glen Aspeslagh, our former students, asked in December 1998 (after they completed our first classroom iteration in genomics), "Okay, so now what are we doing next?" Thank you for asking. It encouraged us to continue, and you two were a major influence in that decision.

Our students who have worked with us in the Wheaton College Genomics Research Group since January 1999 have kept us thinking, planning, doing, and most important learning: Trevor Agnitti, Glen Aspeslagh, Martin Baron, Stephen Benz, Nathan Buggia, Pete Cahalan, Andrea Christoforou, Jonah Cool, Brian Donorfio, Nick Doolittle, Robbie Grossman, Austin Jordan, Neil Kathok, Melissa Kimball, Jon Lister, Sarah Milewski, Nguni Phakela, Patrick Sagui, Jen Tobin, Adam Villa, and Greg Williams.

Kathy Rogers helped with the manuscript preparation. Her tireless work handling the long process and dealing with two sometimes frantic professors was invaluable.

Bill Goldbloom-Bloch (Wheaton College Department of Math/CS) inspired us with his scholarship on Borges, which we found relevant to genomics. Michael Drout (Wheaton College Department of English) discussed many intricate points of medieval manuscript analysis, also entirely relevant to DNA analysis.

We thank Senior Editor Peter Prescott for encouraging us and for helping to shape the proposal for this book. Thanks also go to Alycia Somers and Brian Desmond for attending to all of the details of production. We are grateful to our copyeditor Laura Poole for her meticulous work and for her willingness to take on a project with so many pages of computer code.

We are grateful for the funding of our research and pedagogical efforts from the National Science Foundation (NSF DUE 0340761 and 0126643) and SIGSCE (Special Interest Group in Computer Science Education), as well as research funds from Wheaton College. We are especially thankful for the support of Wheaton alumna Anne Neilson and Provost Susanne Woods.

Mark D. LeBlanc
Betsey Dexter Dyer
Wheaton College, Norton, MA

Preface

This book is an introduction to the Perl programming language for the exploration of biological sequences (DNA and protein) with a focus on DNA for most examples. DNA can be depicted as a linear map of As, Cs, Gs, and Ts, but that map (the direct output of any DNA sequencing project) only hints at the varied contours and crevices, twist and kinks, loops and nodes that comprise the structure and function of the extraordinary double helix. An expedition into that terrain, still mostly unknown, can be approached in two major venues, interdependent with each other.

The first is with the repertoire of techniques that is molecular biology. By those methods, we know genes and have had a good understanding of how they code for proteins since the early 1960s. We also know many short sequences proximal to the genes with specific functions by which genes are regulated.

The second is with the set of computational tools that comprise the emerging fields of genomics (for DNA) and proteomics (for protein). The approach is like large-scale, methodical code-breaking, a search for revealing patterns that suggest meaning. In contrast, even "scaled-up" experiments in molecular biology are piecemeal, one sequence at a time. Wet lab techniques, even if organized and accelerated by their own computers, are limited by the speed of the robotic arms and the incubation times for the myriad reactions. Computational methods are orders of magnitude faster and allow for exhaustive, repetitive, tireless searches and meticulous cataloging and annotation limited only by the imaginations and wills of the programmers.

In any large-scale, comprehensive genomics (or proteomics) project, molecular biology and computational analyses are co-dependent. Computational programs are the lightning-fast map readers, flying across miles of tedious As, Cs, Gs, and Ts and relentlessly collecting and "imaging" every pattern. Molecular biology provides the "ground truth" because many patterns can be easily imagined but prove to be false leads, nothing more than random. Computation has the power to provide hypotheses for molecular biology. For example, "We (the computers) have found an unusual abundance of short sequences 'GTGACGTCAC' upstream of some related genes. We hypothesize that the sequence may be essential in regulating those genes." Molecular biology complements this hypothesis with a mutational study by which subtle point-by-point changes in GTGACGTCAC indeed are found to have significant effects. Then computation completes the loop by allowing more lightning-fast, relentless collecting of every possible permutation of GTGACGTCAC. Maybe more patterns (hypotheses) are to be found, but if not, only seconds or minutes will have been lost!

Perl has been the language of choice when working with, searching through, and identifying patterns in strings of text. DNA and protein sequence analysis is all about strings (for example, linear map of As, Cs, Gs, and Ts) so it is not surprising that Perl has emerged as the programming language of choice for researchers in bioinformatics, genomics, proteomics, and systems biology. As such, Perl is ideally suited to answer questions such as

How can I determine all the three-letter prefixes of every word from a John Donne poem?

How can I determine all the three-letter prefixes of every 7-mer found upstream of my favorite kinesin gene in yeast?

How can I determine all the partial anagrams in a certain word?

How can I search for hydrophobic domains in proteins?

How can I find and count every occurrence of "love" in a psalm?

How can I find and count every occurrence of the putative binding site ACGT upstream of genes in the Krebs cycle?

How can I find all palindromes in the English dictionary?

How can I find all mirror repeats in a genome?

Intriguing and Unique Aspects of This Perl Book

1. Uses an appealing "exploration" metaphor throughout. Biology has returned to an age of discovery, perhaps not seen since the days of nineteenth-century scientific expeditions. For example, Charles Darwin was charged with collecting and cataloging every new organism that he could on his voyage on the *Beagle*. We are there again (biologists and computer scientists as explorers) with billions of bases of uncharted DNA sequences and a great need for simple collecting, cataloging, and annotating. What the intrepid traveler needs is a guide to make the transition into sequence analysis enjoyable and rewarding.

2. Uses a foreign language "phrase book" approach for instant gratification for the novice user. For example using a French phrase book to say, "May I have a croissant, please?" brings the speaker a croissant. Substituting the word *espresso* for *croissant* brings more gratification in the form of espresso. Knowing all of the fine points of the grammar is not necessary to get results.

By that analogy, a biologist (and Perl novice) using our book wishing to search for all direct repeats of length six in a DNA sequence, would say:

```
m/(.)(.)(.)\1\2\3/g
```

And to find mirror repeats would simply substitute:

```
m/(.)(.)(.)\3\2\1/g
```

3. Encourages a "favorite gene" approach by which the user is encouraged to use all possible tools to solve a particular research problem. This is not entirely new for biologists who are accustomed to trying different lab techniques and in some cases using prepackaged software for sequence analysis. What is new is the ease with which biologists can begin making queries specific to their own favorite genes using their own Perl scripts. We strongly believe that a grassroots effort in genomics will solve many of the more interesting problems on DNA function and regulation.

We predict that small labs working on genomic analyses for genes that they already know well will produce some of the more specific and useful results.

4. A serendipitous feature is that linguists venturing into Perl to analyze natural language texts might find this a useful starting manual. In those cases, the English language examples that we use metaphorically would not be metaphors per se but the potential topic for research itself.

What This Book Is Not

This is not a direct route to complete mastery of Perl; this is not a comprehensive treatise on all things Perl; you will not become a wizard or guru unless you are inspired to go much further than the contents of this book. Biologists, we are not urging you to quit your day job and become a programmer. We argue that there is a useful function for a simplified Perl book like this one directly applied to biological sequence analysis.

Who Are the Authors (And How Did They Come to Write This Book)?

Neither of us are Perl gurus or wizards. However both of us are extremely enthusiastic about searching for patterns in DNA sequences and applying it to our own research and teaching students to do the same.

Mark D. LeBlanc, a computer scientist, has written most of his software in C, C++, or Lisp, although Perl now appears more often in his research projects. Like any good programmer, Mark has been trained to learn new languages as needed and as much as needed. Learning Perl was a reasonable next step, but not his initial step in genomics. Like reading poetry, applying Perl to problems in string manipulation led to an appreciation that occurred over time. As an experienced teacher of undergraduate computer science courses at all levels, Mark explains and interprets programming syntax clearly and knows many of the creative classroom tricks for presenting difficult material. He is also a good predictor of the pitfalls that await novice programmers. Mark wrote nearly all of the Perl examples in this book, and Betsey served essentially as the first reader and alpha user of those examples.

Betsey Dexter Dyer, a biologist, is something of a Luddite. She often hand-writes first drafts using a fountain pen that she carries with her everywhere in a special case. On the other hand, she is logical enough to come up with workable algorithms for complex problems, and she communicates well with a diversity of programmers. She has many questions about DNA sequences that seem to be tractable in part via string searching and pattern matching (the strong points of Perl). Some of these questions are being answered in an ongoing research group that she and Mark supervise, which is composed of undergraduate computer science and biology majors who are designing and writing software to decipher some aspects of gene regulation.

However, Betsey's particular contribution to this book is her tendency to take almost nothing for granted. This "phrase book for travelers" approach has a possibility of drawing in a different sort of programmer—an intrepid explorer, very likely a biologist—in need of just enough Perl but without the time or inclination to become a fluent speaker. Having no experience in teaching programming and only one laborious semester's worth of work on C++, and no Perl at all (at the beginning), Betsey had no assumptions or preconceptions about what a book on programming ought to look like and therefore felt free to make up her own formats, many of which were graciously tried and accepted and elaborated on by Mark.

Betsey also wrote almost all of the text set out in boxes, which form a sort of color commentary, and elaboration of examples throughout the book. This is a unique aspect in a programming book and one that we hope may be appealing to a wider audience than that usually attracted to programming.

Who Are You, the Reader?

If you are a novice programmer, perhaps a biologist who is interested in trying Perl (which you've heard many wonderful things about) to do a little pattern matching with DNA (or protein) sequences, this may be a fruitful beginning.

If you are a Perl wizard, perhaps you are quickly leafing through this book out of curiosity, the word *Perl* in the title having caught your eye. Let us know what you think!

If you are already a computer programmer in other languages, this book might be a little too easy for you. However, you may discover within some appealing ways to present Perl to your own students or colleagues, and you may not be familiar yet with all of the sequence applications that we suggest throughout.

If you are a "wordie," an enthusiast of anagrams and other word play (as we are), or if you are a linguist experimenting with string searching to analyze texts, you may enjoy our attempts to mirror language examples with DNA examples. At the very least, you could build your own anagram finder, just for fun.

For the Instructor

All Perl source code contained in this book can be downloaded from the companion website: www.oup.com/us/perlforDNA.

Contents

Perl for Exploring DNA

1 Introduction: Explorations and Indexes

In which the reason for and approach of this book are justified using examples from diverse sources. Enticing reasons for writing your own programs for analyzing DNA sequences: collecting, cataloging, annotating, and indexing as tools for scientific exploration and discovery.

Indexes need not necessarily be dry, and in some cases they form the most interesting portion of a book.

—Wheatley 1878, p. 14

1.1 Darwin's (Hypothetical) Undergraduate Interns

If Charles Darwin had taken a couple of undergraduate interns with him on the *Beagle*, those students would have discovered, described, and cataloged their share of new species. It would have been almost unavoidable, because so few neotropical flora and fauna had been examined and characterized at the time of that voyage. The nineteenth century was rewarding for naturalists with a passion for collecting, pressing, stuffing, pinning, and naming. And then in the twentieth century, such activities fell out of fashion as legitimate occupations of serious scientists. Ernest Rutherford (as quoted by his student, P. M. S. Blackett) summarized the growing importance of the scientific method, especially as defined and refined by physicists: "All science is either physics or stamp collecting" (quoted in Birk 1962, p. 108).

Thus Rutherford, a physicist, dismissed as trivial the work of collecting and cataloging. Admonitions of that sort were taken quite seriously by many scientists in all disciplines as well as directors of funding agencies and editors of journals. E. O. Wilson is among those biologists who lament that attitude about collecting and taxonomy because we have recorded only a fraction of the catalog of organismal diversity on Earth. We scarcely know what lives in the canopies of rainforests or in the oceans or beneath polar ice or the many microbial habitats. A handful of soil is an undiscovered universe of genomes. Taxonomy is by no means a completed and depleted topic.

Therefore it is perhaps ironic that we are experiencing once again an age of exploration and discovery via the old-fashioned activities of collecting and cataloging. This time it is not only organisms but also DNA sequences. Billions of bases of undescribed, essentially unexamined, unanalyzed sequences are being stockpiled, the numbers growing daily, in the convenient databases of the National Center for Biotechnology Information (NCBI) and other major public databases. The sequences are available for anyone, expert or amateur, to commence any study, on any scale, with almost any purpose. As it happens, collecting, sorting, annotating, and describing DNA sequences are high on the list of what needs to be done. Even the most obvious

---

Box 1.1 The Fine Print

....

Some of the things that you will get little or none of in this book: You will not be given really explicit instructions for the standard use of preexisting software. Our goal is that you might dare to write your own or participate actively in a group that is writing its own. Yet we are not necessarily trying to produce professional programmers from novices. It is almost more important, for the way in which much of science is done, to become a versatile, functional member of an interdisciplinary group. That might very well involve learning some Perl, while at the same time not quitting your day job as a biologist or chemist.

You will not be advised too strongly about what sorts of analyses you should be doing. For example, we won't tell you that you ought to be building phylogenetic trees or studying introns. We leave that up to you, assuming that you have lots of good ideas and are seeking creative ways to implement them. Perhaps we stand a chance of the book not going out of date too quickly as one or another type of analysis goes in or out of fashion. We do, however, have our own personal preferences about what sorts of queries and analyses are interesting, and these will be scattered throughout the book as models. One example that will soon become apparent is that we think very highly of concordance or lexicon building as a first step to deciphering sequences. Even if you don't agree with us on that, many of the models we will show are good starting points for your own personal interests in writing software.

We will not encourage you to write your own software when perfectly good software might already exist (and is accessible) for your particular query. The NCBI is a rich source of computational tools, including BLAST, and only gets better every year. Check there often to see what is new and usable and time-saving for you.

---

analyses, such as "gene finding," are available as research topics because gene finding, especially in eukaryotes (organisms with complex cells), is still an imperfect science. Among the complications are abundant pseudogenes (highly mutated and no longer functional) and the many splice forms of genes by which one sequence might result in many diverse products. Furthermore, many genes have been transferred horizontally from one species to another (as opposed to vertically in a tree hierarchy), thus complicating the building of sequence phylogenies. The enormous task of assigning names and functions to genes and of grouping genes into logical gene families and super families (essentially taxonomy) lags far behind the accumulation of data from DNA sequencing projects.

The greatest uncharted territories are the sequences between genes (intergenic sequences) where it is possible to venture into long stretches of DNA never before examined. Within those intergenic regions, comprising the majority of eukaryotic genomic sequences, genes are controlled by mechanisms not yet completely deciphered. Maybe DNA sequence analysis is a form of stamp collecting, but it is also an exciting time in the history of science to have so much raw data, so potentially laden with meaning, awaiting exploration on almost any level.

1.2 How Much DNA Is There in a Genome (Using the Human Genome as the Example)?

Nearly every human cell has two copies of a genome of 3 billion base pairs. So that makes a total of 6 billion base pairs per cell. Depending on how tightly the DNA is coiled, there are about 4,000 base pairs per micrometer (micron) of DNA. A little calculation reveals that every cell contains the astounding length of 1,500,000 micrometers of DNA. That number can be made even more vivid by converting it to meters, cutting a string of that length, tucking it into your pocket and slowly pulling the string out, after having elicited guesses from onlookers as to the length. The string is 1.5 meters! How such an enormous polymer (no matter how thin) could be stuffed into a nucleus of about 1 micron diameter and remain functional is a topic of ongoing research.

Now consider that there are about 10 trillion cells in an adult human being. (Estimates vary. It is difficult to count.) That means that we carry around about 15 trillion meters of DNA, a number almost impossible to imagine unless converted into other units, although even representing it as 15 billion kilometers doesn't quite do it. Note that the sun is about 150 million kilometers from Earth. That means we each have enough DNA to string out into space 100 times the distance to the sun.

By the way, most of the cells we are hauling around are not the 10 trillion of our own but rather our load of symbiotic bacteria, 100 trillion of them. And each of those bacteria has its own genome and so on.

For more calculations about DNA, see chapter 5, in which the numbers are about the printing of letters A, C, G, T to fill the right number of books to contain the human genome.

How Many Genes Are There in the Human Genome?

This is still a contentious matter partly because of the difficulty of defining the word *gene* and partly for political reasons. The definitional problem centers on the extraordinarily flexible way that eukaryotes use their genes to get numerous protein products out of one sequence. Exons (functional sections of genes) may be spliced together, resulting in many combinatoric possibilities. If we tally up messenger RNAs or proteins in a cell, we should get far more than the number of genes. Furthermore, different gene predicting programs stumble in different ways over introns (the noncoding parts of genes) and pseudogenes (genes that are no longer functional due to accumulated mutations). The political problem centers around the commercial interests in "owning" or at least controlling parts of the genome. In this context genes can be viewed as marketable units, each representing a potential function, cure, and profit.

GeneSweep was established as a friendly wager at the Cold Spring Harbor Genome Meeting in 2000. Bets ($1 the first year, $5 the next, and $20 after that) were recorded in a lab notebook. Participants submitted estimates of the number of human genes. These estimates from experts in the field ranged from 26,000 to 150,000, which gives an idea of the challenges of gene prediction. Finally in

2003 the human genome was declared "done" (at least from the perspective of the popular press), and the number of genes was announced as being in the range of (a shocking) 24,000. Clearly complexity of function and structure is not correlated with the number of genes! Note that *Caenorhabditis elegans*, a model worm of about 1,000 cells and 1 millimeter in length has about 20,000 genes. (The winner of GeneSweep was Lee Rowan of the Institute of Biology in Seattle with a guess of 25,947.) However, most genome biologists agree that the number is far from settled and considerable work is still needed in the area of gene finding software.

What About the Enormous Spaces between the Genes?

With only about 24,000 genes, lightly sprinkled throughout 3 billion base pairs, the spaces between the genes, the intergenic sequences, are enormous. Let's roughly estimate the length of a gene as 10,000 base pairs. (Gene lengths vary widely.) That means that genes occupy approximately 240 million base pairs, which is only about 12 per cent of the genome. So take that 1.5-meter length of string from your pocket and pinch off about 18 centimeters. The rest of the string represents the intergenic regions. Because it is now becoming pretty clear that we cannot attribute our complexity to our quantity of genes, we really must focus on the spaces between.

What About the Intergenic Sequences?

They aren't junk. Those sequences are where most of gene regulation is being encoded, resulting in the crucial and subtle differences in how our limited repertoire of genes is being used. It is where "complexity" is encoded.

However, the metaphor of referring to DNA, especially intergenic DNA that has not yet been deciphered as "junk," is so simple and visual and easy for the general public that it is used over and over again by popular science writers. Unfortunately, the phrase "junk DNA" is certain to be with us for quite a while yet. No matter how many new studies come out dispelling the idea of junk DNA, by patiently deciphering and finding meaning in yet another supposedly useless piece of it, reporting by the popular press continues to be biased. Each new meaningful bit of sequence is dismissed as an odd exception. For example, long repeats of intergenic DNA (microsatellites) have a history of being declared and dismissed as junk. However (and this is just one of many examples), the length of a certain microsatellite sequence distinguishes two species of voles. A longer microsatellite near a gene for hormone control seems to confer greater paternal care of off-spring and thus a different family structure in the mole nest (see Hammock and Young 2005).

Perhaps one problem is that such examples viewed one at a time do seem to be trivial and even anomalous. Vole paternal care might well be an exceptional case and the rest of intergenic DNA still is junk! Such examples viewed not one at a time but by the thousands will perhaps lay the myth to rest some day. It isn't junk; it just hasn't been examined and deciphered yet.

1.3 Computers as Serendipity Machines

Serendipity, n., the faculty of making happy and unexpected discoveries by accident (*Oxford English Dictionary*).

The new and essential focus on collection, storage, and analysis of huge quantities of sequence data means that there are some "new" ways of doing experiments, including forming hypotheses and following through on them. Actually these new ways date from the origins of biology in natural history. David Green in *The Serendipity Machine* refers to fortunate accidents that occur during exploration and discovery: "For most of human history, information was scarce and expensive; now it is abundant and cheap" (p. 5). And that has changed some ways of approaching scientific problems. The larger the data set and the more quickly and freely it can be explored or mined, the more likely it is that new combinations, connections, and relationships will be found among the data elements. Computers are "serendipity machines," essentially engines for creating serendipitous discoveries, says Green. It is a legitimate (albeit still controversial) scientific activity to seek patterns first (often at lightning speed via computers) and contemplate and hypothesize later! "This is a relatively new way of doing research which came to the fore during the 1970s as the power of computers increased. However many older scientists were (and still are) uncomfortable with the idea of deliberately setting out to make scientific discoveries without knowing what you are looking for. This is surprising because science has always exploited serendipity" (p. 89).

Indeed, Darwin on the *Beagle* was relatively unencumbered by hypotheses. Collecting and cataloging could have been ends in themselves, the specimens stored safely in museums. It was only later back in England that Darwin reexamined his specimens, observed the patterns and distributions, and proposed an origin of species by natural selection.

Green points out that one reason that large data sets (or large collections such as Darwin's) provide so many opportunities for accidental discovery is because there are so many possible combinations of the individual items. No matter how a database is designed (or a collection arranged), unanticipated patterns will emerge. The larger the database, the greater the possibility of serendipitous discovery. Indeed, the arrangements of data ought to be frequently examined as new items arrive daily because old hierarchies and taxonomies might no longer fit the new information. Green calls this the "platypus effect" after the temporary chaos that occurred when duck-billed platypuses were discovered in Australia and could not be neatly cataloged with the other mammals until the classification was adjusted. Hierarchies and classifications are no more than hypotheses. Current practices in storing, sorting, and annotating DNA may need to be reexamined as more information pours in. Some of the new sequences may have platypus-like effects on the databases.

1.4 Indexing as a Research Activity, Especially in Bioinformatics: Indexes as Discovery Tools

Bioinformatics is presently in its glorious indexing phase, and we do not know how long this phase will last except that sequences keep rolling in and so far annotation has not kept pace.

At first glance, building an index would seem to be a mundane and even tedious chore that necessarily follows the writing of a book and not much more. Concordances (essentially an index of words and their locations rather than subjects) and even the more elaborately detailed versions of indexes, such as catalogs do not seem significantly more exciting. However the same sorts of decisions about what to index and how to place an indexed item into a hierarchy and whether to cross-index and add annotation are all common practices of DNA annotation.

The act of indexing (or concordance building) essentially allows you, the curious researcher, to get first dibs on virgin genomic sequences! It allows you to make your mark or impression on the raw data, pulling it into some semblance of being informative. In fact, until the first steps in indexing are taken, genomic sequences might as well have been generated randomly, so little information is revealed. You are indeed building an index or catalog or concordance when you

1. Annotate any part of a genome—seeking and recording genes, introns, promoters, and other features. Objectivity is of course a goal, but subjectivity often guides the decisions as to what gets emphasized.
2. Look for and record binding motifs, protein domains, and restriction sites, including motifs of (as yet) unknown function—such as small inverted repeats in promoter regions.
3. Organize sequences according to metabolic pathways or co-regulated genes.
4. Look for and record transposons, viruses, pseudogenes, and other intriguing genomic sequences.
5. Attempt to decipher the promoters of suites of co-regulated genes in a pathway.
6. Perform all sorts of counts of oligonucleotides—such as A, C, G, T ratios, dimers, trimers, tetramers, and so on.
7. Do any of the above while comparing and contrasting genomes.
8. Make phylogenetic trees. The tree itself is an informative, nonlinear index!
9. Create graphic map-like representations of sequence information. Maps, too, are nonlinear indexes.
10. Produce software that filters or mines data that is essentially index building, complete with all of the best guesses necessary as to what is worth collecting.

All of the above are hypotheses or hypothesis generators, subject to change and rich with potential discoveries. What you choose to include and in what hierarchies and categories and relationships represent your best guess as to what is meaningful and appropriate. What you omit or exclude may be just as revealing.

Commenting on the subject of indexing books, A. S. Byatt (2001) notes that some of the late entries to an index are made only after the realization that a particular topic is relevant. An example is the Dewey decimal system (essentially an index of books), which was invented long before computers. Therefore there wasn't much space in the system for books on programming. Where are they now? They are within 003 to 006, tucked in between books about books (002) and bibliographies (010), a sort of best guess at the time as to where such a topic ought to reside. Such guesses (or hypotheses) about relevance, connections, and hierarchies may be worth considering for the organized storage of DNA sequences as well. Making the decisions as to what is informational and what is random is one of the most daunting and essential, challenging, and exciting tasks of a bioinformaticist.

1.5 Why Write Your Own Programs for DNA Analysis?

Bioinformatics is still an immature field with many as yet unknown possibilities for new methods and equipment. There is plenty of room for inventing new procedures. Most sciences started off that way. Then as the various fields matured, equipment and procedures "fossilized," and you can assume that researchers are no longer building their own. For example, chemists used to blow much of their own specialized glassware, and chemistry departments often still maintain glass-blowing rooms, to the wonder of undergraduates who simply use glassware off a shelf in the stock room. Molecular biologists used to build their own gel boxes with supplies from the hardware store (clamps, panes of glass, and wires). Protein sequencing projects at one time began with building your own sequencer. Furthermore, there is a long tradition in science of using modular parts of equipment to invent new equipment. That means that many science departments still have drawers and cabinets full of glass connectors and odd bits of tubing—relics of a more inventive past. Bioinformatics is still very much at that invention stage. The metaphoric drawers and cabinets of stuff are still in use. Although it is true that some bioinformatics software can be taken and applied directly from the shelf, much of it still remains to be written. The field is appealingly undeveloped (compared to other scientific fields) and open for fascinating, new approaches.

Note that none of the following list of Sixteen Reasons to Write Your Own Code precludes:

1. Using preexisting software. For example, there is no need to rewrite a BLAST-like program for yourself if BLAST is what you need. Being able to program makes it easier to understand preexisting software—how it works and what its limitations might be.
2. Getting professional programmers to write complicated software with thousands of lines of code, with intricacies that takes years of experience.

Sixteen Reasons to Write Your Own Code

1. You have some really creative (so far unique) questions about genome structure and function and you don't want to be held back by software.
2. You like tinkering with things, taking them apart, putting them back together.
3. You (like an old-time explorer) are adventurous and even fearless!
4. Writing software is actually a hypothesis-building activity—because of the decisions and priorities that have to be made at every step.
5. After using a commercial software package, you came up with additional queries that went beyond the capabilities of the software.
6. You are in a working group that includes programmers and you'd like to communicate with them a little better. You want a better idea of what the possibilities are and how quickly they could happen.
7. You might find that you really enjoy building software. In finding that out, you will get an even better rapport and empathy with programmers, further improving communication and team synergy.
8. You want to get a better idea of when you need to reinvent the wheel and when preexisting software is enough or tweakable.

9. You'd like to experiment with programming to see if you like it well enough to shift your career into that direction.

10. The major producers of sequence software are the government and industry. Their research questions may be motivated differently than yours. Their questions may have more to do with quick, reliable profits. You might want to do more open-ended exploration based on curiosity.

11. Some of the major software packages for bioinformatics are not subtle enough for the types of questions you want to ask. The results are too "vanilla."

12. Some of the public software is too expensive.

13. Some software is proprietary, built at considerable expense by biotech companies. You might hear about it, but you won't be able to use it.

14. Some software gives results slowly, one item at a time, forcing you to go in again and again to put together a complete search.

15. Some of the public software is not user friendly (or at least it hasn't been very friendly to you).

16. You are using a nice piece of software but are occasionally plagued with thoughts of "what is this software *not* showing me? What decisions did the programmer make to leave certain things out?"

Box 1.2 Finding Words and Motifs with Your Word Processor

Why not commence your first endeavor in sequence analysis right now with your word processor? You have Find and Replace functions in the edit menu. You also probably have the capability of doing a wildcard search with an asterisk, as in "wi*dcard." This feature might not pop up immediately but may be revealed with a little sleuthing using online Help. Remember, you also have the capability of counting characters, with and without spaces, as well as counting words, lines, and paragraphs. Some sequence analysis software begins with these basics. Some of the first programs you write may not do much more than what these built-in functions are doing.

Using your word processor, create a new file and type in a sequence of As, Cs, Gs, and Ts like this:

acttgcaaatgcgcatgttaaaccccctgtcagatgacccaaattacgaaatgtgtcgcgtatgacgaattgcgc

By using Find, you immediately and effortlessly can locate each instance of "tt" or "gcat" or any other motif of interest. Try a wildcard with "c*aat" for a fuzzy search. If you had several pages of sequence to search through and merely wanted to discover and underline every position of "atg," some of which could be serving as start codons at the beginnings of genes, your word processor might suffice. Use Replace to make all of the instances of "atg" bold, capitals, or underlined.

acttgcaa**atg**cgc**atg**ttaaaccccctgtcag**atg**acccaaattacgaa**atg**tgtcgcgt**atg**acgaattgcgc

In addition to playing with the features on your word processor, you may want to open your Internet browser. Online search engines typically have some versatile features, including Boolean searching and wildcards. We would like to point out that simply typing in a DNA string (or protein) of sufficient length, such as "gtgcacc," might quickly reveal some research or annotation being done on that sequence. We have noticed that although short trimers and tetramers that tend to get used for acronyms are not good search subjects, longer motifs of interest are definitely worth trying. We have used similar searches at PubMed (MedLine) to find scientific abstracts that feature particular sequences of DNA or proteins.

Box 1.3 Annotating Sequences by Hand

One of the conventions for naming short, meaningful sequences of DNA is to call them boxes as in "TATA box," "CAAT box," and "homeobox." This dates from the time when DNA analysis was mostly done by hand with colored pencils or pens. Sequences of potential interest literally would have a box drawn around them. In 1975 David Pribnow reported in the *Proceedings of the National Academy of Sciences* on the significance of TATAAT as a binding site in the promoter region of some prokaryotes. Shortly thereafter, colleagues began to refer to that sequence as a Pribnow box and the term *box* entered common parlance for any sequence of interest. Some biochemists use a dry sense of humor in naming sequence elements as in CAAT box (a binding site of eukaryotic promoters), evocative of a box of kitty litter.

Colored pencils still have a role in DNA sequence analysis. If you are writing a program to execute a search for some pattern or motif in a sequence, you need to start with a very small test file that you have marked up, in advance, to locate your sequences of interest. If your program succeeds in finding the same sequences, then perhaps you are ready for larger files. Otherwise, there may be some error in your program. Part of the ground truth of programming is checking enough results by hand that you can trust what your machine is doing for you.

2 Essential First Steps: Installing Perl Plus a Programming Environment

In which Mac OS X users and Windows users alike will eventually get their respective machines set up for programming in Perl.

When you have eliminated the impossible, whatever remains, however improbable, must be the truth.

—Doyle 1890

Our goal for this chapter is to keep you going with as much encouragement and support as possible. Getting through this chapter with success will make the activities of the subsequent chapters possible. The instructions here are the sine qua non of the rest of the book because they tell you how to get Perl up and running on your own computer.

Two major hurdles for us in writing this chapter were:

1. The Web sites from which Perl and a programming environment for Perl may be obtained, often for free, are moving targets. Overly explicit instructions will quickly date this book and frustrate users. We hope that you will be able to soldier through extra Web pages that might pop up during the installation, which were unforeseen by us. Note that we are focusing on free, open source download sites. However, there are programming environments that can be purchased; if you are in an academic or industrial setting, you may already be licensed to use one of these. It is worth checking with your systems administrator.

2. Our readers have different operating systems. You may be using Windows, Mac OS X, or the Linux operating system. Therefore we have multiple sets of instructions and links on our Web site accompanying this book. They are presented briefly here, but for full details, go to the site. For various historical reasons, as of this writing, Perl is included with Mac OS X and Linux but not with Windows. Therefore the Windows folks will have to trudge through a lengthier protocol, rendered a bit friendlier by us.

Beginning programmers may be puzzled as to why you need to install two things, Perl *and* a programming environment for Perl. Why not just Perl itself? Indeed, Mac OS X and Linux users may be aware that Perl comes bundled already with the operating system. That is one reason that Perl installation instructions for Mac OS X and Linux are so brief. However, almost all programmers prefer to write software in a comfortable environment. Once installed, Perl resides in your computer as an enormous set of useful commands and functions as well as the grammar and syntax by which they are used. A programming environment adds an attractive, color-coded font and an easy way to run programs. It may also highlight incorrect spelling and syntax to help you spot errors in your programs. While some intrepid experts in a hurry may run Perl on a stark, black screen, many programmers prefer the extras of a pleasant environment.

We have made the online instructions for installing Perl and a programming environment somewhat verbose and full of commentary. Depending on your level of confidence, you may not need the extra wording and detailed instructions. Everyone else, read closely, especially if you are acting as your own system administrator.

Visit the following URL for directions on installing the Perl interpreter (engine) and a programming environment.

www.oup.com/us/perlforDNA

Click the appropriate link depending on whether you are using a Macintosh (Mac OS X) or a Windows computer.

STEP 1: Installing perl and a programming environment

for Mac OS X

Download PDF Instructions

for Windows®

Download PDF Instructions

Box 2.1 Two Reasons to Skip This Chapter

1. You already have Perl plus a comfortable programming environment for Perl loaded onto your computer. This might well be the case in certain academic and industrial settings. Congratulations! Skip to the next chapter. However, we will repeat our warning to Mac OS X users. Perl comes bundled with your operating system, but the pleasant programming environment does not. You will have to install that. Do not skip to the next chapter.
2. Down the hall is a team of IT folks or the help desk or a computer science professor, with whom you have a good working relationship. They are somehow not too busy to help you load up Perl and a programming environment. Feel free to walk down the hall! Please note, however, that these are likely to be very techy folks, accustomed to rapid typing of terse commands and lots of shortcuts. If you would like to learn something about the installation, you may have to ask them to slow down and explain what they are doing. They may have suggestions for where to go to download what you need, especially if Perl is one of their favorite languages. On the other hand, they may not be familiar with where to get Perl, so be sure to show them the links in this chapter.

Box 2.2 Using This Chapter to Forge a New Relationship

If you have access to an IT group or a help desk or a Computer Science (CS) department, you may find that your questions about installing Perl will open up an entirely new and positive set of interactions. Your computer-literate colleagues are probably tired of going around all day unfreezing screens and jiggling or reattaching cables. Tell them that you are installing Perl and are going to learn to program in it. They will be intrigued! It may turn out that Perl is their favorite language. They will understand immediately why it is applicable to searching strings of DNA and may have all sorts of good suggestions as you undertake the next few chapters. One of our favorite pieces of advice for biologists embarking on the rudiments of Perl is to venture down the CS hallway or across campus to the CS building. Do not underestimate the level of interest and potential collaboration that may result.

3 Your First Perl Programs with a Focus on String Analysis

In which you will get through "hello DNA Land" and proceed to the practical fun of using Perl in one of its great strengths—the analysis of strings such as DNA sequences.

Perl is remarkably good for slicing, dicing, twisting, wringing, smoothing, summarizing, and otherwise mangling text.

—Stein 2004

The Perl programming language (Practical Extraction and Reporting Language) was not invented specifically for DNA sequence analysis, but it might as well have been. The parallels are many between the analysis of texts in natural languages and the analysis of DNA strings or sequences. For this chapter in particular, there is a great deal of material that you really should get through. String analysis is the essence of DNA analysis. Therefore, we are cutting this introduction short and urging you to read and work through these examples as you sit at your computer. Take a break from programming later and read some of the text boxes in this chapter on topics such as the amazing Larry Wall, inventor of Perl and the supportive Perl community. Meanwhile, let's move on to "hello DNA Land."

....

If you have skimmed eagerly ahead to this chapter, failing to stop at chapter 2, go back! You cannot begin to write programs until you have installed Perl and a Perl programming environment on your computer.

....

3.1 My First Perl Program

A first program is a special event. It serves as a quick test to ensure that your programming environment is working properly, and it also begins a quest to learn the syntax and semantics of a language that can make a computer do your bidding.

A first Perl program is shown below. Assuming you have successfully installed a Perl programming environment (see chapter 2), use your Perl programming environment to create a new Perl file called helloDNA.pl. You should begin the practice of using an appropriate file name, the name providing a hint at the task that the program will perform and ending with the Perl (.pl) file extension.

Good Practices

When creating file names, we recommend that you concatenate multiple words, use lowercase letters except when starting new words, and avoid spaces. Examples of good filenames include `helloDNA.pl`, `findMotif.pl`, and `fuzzyGeneFinder.pl`.

`helloDNA.pl`

```
#!/usr/bin/perl

use strict;
use warnings;

# This is my first Perl program

print "hello DNA Land";
```

This Perl program prints the message "hello DNA Land" to your computer screen. To run this Perl program, find the Run menu in your programming environment and select it. (Note: in some environments it is necessary to save the program before running it.) After you run this program in your Perl programming environment, you should see an output window containing your message. An example output window is shown below.

Let's quickly discuss the five instructions in this initial Perl program. The Perl statements appear on the left; the right-hand side is annotation and is not part of the Perl program.

`#!/usr/bin/perl`	*Lists the typical location of the perl interpreter ("engine") and should be included as the initial line in every Perl program.*
`use strict;` `use warnings;`	*These two lines help to discover potential errors in your Perl. We recommend that you always include these lines.*

# This is my first Perl program	A "comment" line. The text after the hash (#) symbol is an English explanation or comment, not a Perl instruction.
print "hello DNA Land";	An instruction to print the message between the double quotes to an output window on your computer screen. Note that each instruction or statement in Perl is terminated by a semicolon (;).

As shown, the print statement and other Perl statements introduced later are terminated with a semicolon (;). Although not usually recommended, a single statement may appear on multiple lines, that is, the new line character at the end of a line does not terminate a statement; a statement is terminated at the semicolon. For example, the following is one legal statement appearing on two lines.

```
print
   "hello DNA Land";     # one Perl statement on two lines
                         # the semicolon (;) terminates
                         # the print statement
```

Good Practices

Inserting comments into your Perl programs provides a human reader with a natural language explanation of the program and is one of the first practices of good programming. In the Perl examples that appear throughout this book, we use comments wisely, and we encourage a liberal use of comments throughout your own Perl programs.

This is a subtle difference between double quotes ("...") and single quotes ('...') that we will cover shortly. For now, use double quotes to surround the message that you want Perl to print to your output window.

Try This Now

1. Programming is a hands-on activity. At this point, we are assuming that you have opened your Perl programming environment and created a new file named helloDNA.pl. If you haven't tried it yet, type in the "hello DNA Land" program exactly as shown and then run this program. (Note: in some environments you may need to save first.)
2. Change the message between the double quotes and rerun your program.

Box 3.1 The Perl Community

....

The fact that there is something called "the Perl community" says much about the open and congenial nature of the Perl language and its users. To get a glimpse of it, you could join one of their online email lists and read some posts, essentially listening in. For example, look for the Perl Mongers site that has links to many regional groups. The dialogues will likely be loaded with jargon and queries about especially difficult problems. For example, " I bet you can work around that by saving STDOUT, reopening it on IO::Tee and having IO::Tee output to the file and the saved STDOUT." Even when they are planning to go out for beers, the exchange of posts can sound as though they are planning an especially technical expedition to the peak of Mount Everest.

Should you make amateur queries on a professional Perl discussion group? Use some caution. The answer could be overly technical and even a bit sarcastic. It might be better to go to one of the many online Perl tutorials and do some look ups.

Keep an eye on the discussion lists for speakers coming into town. Some of the major innovators of Perl enjoy stopping in at a "technical meeting" and either giving a talk or just conversing about Perl. You may never get involved enough to become part of the main dialogue, but you may get something out of attending one of the talks. If you are feeling even more intrepid you can find out when the next YAP (Yet Another Perl conference) is going to be held and attend.

The Perl community provides an important service for all of us, no matter how amateur our use of Perl might be. The community has stockpiled and annotated lots of useful code and suggestions for putting together programs. It is easy to make queries and look up bits of code that are actually called look ups. Make queries for code at Perldoc.com; for many different useful links go to Perl.com.

3.2 Controlling Output

Controlling how the results from your Perl program appear in the output window is an important part of programming. Although your program might take minutes or even hours to search and compute, the output from your `print` statements are your only source of knowing how your program is proceeding and if your final results are correct. It is always worth the time to write neat and explicit messages to print from your programs. Later on, as you write more complicated programs, the practice of writing good `print` statements will serve you well.

Each `print` statement literally outputs the message between the double quotes, but unless told otherwise, each `print` statement continues printing wherever the previous message ended. Put differently, a series of `print` statements will continue to print on the same line even if they are on separate lines of code, as shown in the file `testPrint.pl` and the associated output.

testPrint.pl

```
#!/usr/bin/perl

use strict;
use warnings;

print "one";
print "two";
print "three";
```

To force a subsequent line of output to begin on a new line, you can output a new line character explicitly by ending your message with \n. The backslash-n combination is referred to as an escape sequence. The backslash informs the print statement that the next character is not to be interpreted as the normal alphabetic letter n but rather as the special character that forces the output to the next (new) line.

If we add \n to the end of each of the Perl print statements,

```
print "one \n";
print "two \n";
print "three \n";
```

we will produce the following output

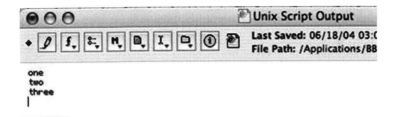

Each \n combination forces any proceeding output to begin on the next line. It is possible to use more than one \n in a print statement, as shown in the next example. To ensure that you understand both the absence and use of \n, carefully read the Perl print statements in the next example and compare your expected output with the actual output.

```perl
#!/usr/bin/perl

use strict;
use warnings;

# A favorite quote
# Reminds me of the circular genome of prokaryotes

print "The number of pages in this book \n";
print "is no more or less \n";
print "than infinite. \nNone is the first page, \n";
print "none the last. \n";

print "\nFrom the short story \n";
print "The Book of Sand (El libro de arena) \n";
print "by Jorge Luis Borges, \n";
print "an Argentine poet, essayist, and short-story writer. \n";
```

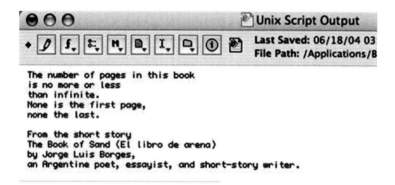

Good Practices

A good rule of thumb is to write short, succinct `print` statements; a long message is best done over multiple `print` statements rather than one long message.

Another useful escape sequence is the tab character (\t). In later programs, we'll want to produce tab-delimited output so we can perform subsequent analyses and formatting with a spreadsheet program. Printing tab characters between column headings and data values ensures that your Perl output will automatically place tab-delimited data in their own columns when opened by a spreadsheet program such as Excel. A sample `print` statement that produces tab-delimited headings for a table of data is shown next.

```perl
print "A, C, G, T ratios per Organism\n\n";
print "Genus\tSpecies\tA\tC\tG\tT\tother";
```

Because the Perl print statement outputs all text between the double quotes, if you want to include double quotes (") in your output, you will have to precede the double quote you want printed with a backslash, that is, \".

```
print "Welcome to \"DNA Land\"!!!. ";
```

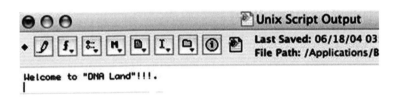

The escape characters are summarized below.

Escape characters	If you want your output
\n	forced to the next line
\n\n	double-spaced
\t	tab-delimited
\"	to actually print a double quote

Going Back for More

1. Practice with various combinations of the new line character (\n). For example, print the chemical names of the four nucleotides, one per line (adenine, cytosine, guanine, thymine).
2. Alter the program to print two nucleotide names per line, each line containing the names of the nucleotides that base pair one with the other.
3. Change the program that prints the quote from Borges to print a different quote that applies to genomics. Be creative!
4. Start a new program to create a picture of the double helix, including matching base pairs. The basic set of characters found on a computer keyboard is known as the ASCII (American Standard Code for Information Interchange) character

set. Creating pictures with only ASCII characters is known as ASCII art. Email smilies or emoticons are a primitive form of ASCII art, for example "happy" :) and "surprised" :o. Search the Web for "ASCII art" to get ideas. You can find examples of ASCII art, copy the picture into your program, and then insert the correct double quotes around each line.

....

Programming in a language like Perl gives you the power to enable computers to do the following two things and do them well: remember (store) things in memory and process (compute) new results. The remainder of this chapter covers how you can store sequences of DNA and other types of strings in the computer's memory and then begin to compute new results using built-in functions that are ready to work on your DNA sequences and numerical values.

Box 3.2 What Is the Best Language?

....

Here is a good conversational topic for any group of computer science professors. "What is the best programming language with which to introduce undergraduates to the computer science major?" The conversation may even become a bit contentious as the virtues of Java, C++, and Scheme are debated. Then it may become a bit sentimental as folks reminisce about Pascal. However, even if there is a Perl guru in your midst, he or she will probably not insist on Perl as the perfect starter language. Why? Well, Perl is missing some of the higher unifying theory of what is considered good programming practice. It is not an ideal way to introduce the general principles, because Perl is a bit careless and cavalier about those principles. Here and there, Perl seems to have been put together by a very creative committee (the "Perl community," headed by Larry Wall), eager to insert as many great features as possible, regardless of the big pedagogical picture. Perl does not set a mature tone. Perl is a renegade. It is fun for outlaw geeks, but creating outlaw geeks is not the goal of undergraduate computer science departments.

Now start another conversation with that same group of computer science colleagues. Perhaps do this the next day, because they'll be exhausted after settling the issues in paragraph one. Only this time your question is, "What is the best language for analyzing and matching strings?" Here the answer is likely to be more concordant—"Perl, of course"—even from those who do not themselves use Perl. And furthermore (it will be agreed), Perl is friendly to regular expressions, which means that all sorts of awkward fuzzy string matches will actually be facilitated. Those programmers who are trying to do (or teach) some string-matching in Java or C++, are facing some awkward obstacles in their syntax. And they probably aren't doing regular expressions at all; it's just too messy.

What is the difference between the two questions? In the first case, in which a new programmer is being guided and developed, the big principles ought to hold true, no matter what the applications. Thus the first programming course is laid out like a set of architectural blueprints for future construction. On the other hand, some of us would like to start matching DNA sequences immediately with the friendliest tool for the job. Are we training to be an architect or building a picnic table? We would suggest that for encouraging beginners (either biologists or computer scientists) to get right into the most interesting problems of DNA sequence analysis, Perl is the clear choice. Furthermore, as we reiterate throughout this book, all the myriad functions of DNA have not been worked out. Much remains to be decoded in the billions of base pairs in storage at NCBI. There is plenty of room in the field of DNA sequence analysis for creative questions and approaches. Perl lets you implement them fast and efficiently and rewardingly—that might be reason enough.

3.3 Variables

Computers have an impressive memory. Whether remembering a particular nucleotide or an entire genome, all programming languages (including Perl) provide programmers places in memory to store or "structure their data," commonly referred to as "data structures." Data structures come in various sizes and shapes, the most basic being a variable. We will use variables in Perl programs to store and retrieve many different types of data, including DNA sequences, English phrases, and numerical values. These simple values are referred to as *scalars* in Perl.

Because we are suggesting that you always include use strict at the start of your programs, you must first declare each variable with the Perl keyword **my**. Keywords in Perl are special words that perform a specific function. All scalar variable names in Perl begin with a dollar sign ($). A sample variable declaration is shown here.

```
my $bindingMotif; # a variable to hold a short sequence
```

At this point, the variable called $bindingMotif is empty. That is, declaring a variable only sets the name of your variable. No particular sequence is associated with or stored in the variable named $bindingMotif at this point.

....

Good Practices

Comments can be positioned on their own line or at the end of a Perl instruction. If the comment is a summary statement and takes more than one full line, we recommend placing those comments together in a left-justified block. It is also a nice practice to use a comment line that separates your comments into a box and to use blank comment lines to keep your comments looking neat.

```
#-----------------------------------------------------------
# This program tests my knowledge of string variables and
# using Perl's built-in string functions and operators.
#
# Programmer: Tina Tubulin
#
# Date: January 8, 2006
#-----------------------------------------------------------
```

When declaring variables, we recommend that you place your comment on the same line as the variable declaration, lining up each comment with subsequent variable declaration lines. This makes a nice indexing of sorts of your variables with each variable annotated with a note about its role in your program.

```
my $genusName;   # genus name of organism
my $speciesName; # species name of organism
my $genomeSize;  # total size of genome (bp)
```

All comments in Perl are optional, but we encourage the liberal use of comments. You should consider the comment as an English paraphrase of your Perl instruction. Although we won't insist that you write a comment for *every* line of Perl, later sections will introduce commenting conventions for you to follow so your programs meet a professional standard (and make you look like you've been programming Perl for years!).

One way to store a sequence or value in a variable is to use the assignment (=) operator, as shown below.

```
$bindingMotif = "ACGT"; # store "ACGT" into the variable
```

It is sometimes helpful to think of variables as boxes. A variable (box) has a name, the name serving as a reference to a memory location in the computer's internal memory. The variable (box) stores the DNA sequence or numerical value that is assigned (=) to it.

```
my $bindingMotif; # a variable to hold a short sequence
```

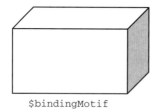

$bindingMotif

```
$bindingMotif = "ACGT"; # store "ACGT" into the variable
```

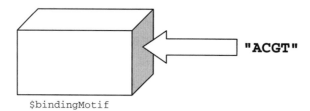

"ACGT"

$bindingMotif

The assignment statement

```
$bindingMotif = "ACGT";
```

should be read as: "Assign the sequence ACGT into the variable $bindingMotif." (Avoid saying "$bindingMotif is equal to ACGT" because this would be a true statement only *after* the statement finishes but would miss the relevant action of the assignment that is taking place.)

....

Good Practices

Because you will use a variable throughout your own program, you will be well served to follow these tips when deciding what names to give your variables.

1. Although Perl allows other combinations, we recommend for good style that your Perl scalar variable names begin with a required $ followed by one letter followed by any combination of letters, numbers, or underscores (_).
2. Use a variable name that immediately brings to mind the contents that will be held in that variable. Good names include: $averageMotifLength, $organismName, $GCratio. Poor names include: $abc, $total (total of what?), inside jokes such as $foobar, and generic one-letter algebraic variable names such as $x and $y.
3. Use lowercase letters for your variable names (see below for the exception).
4. When appropriate, use more than one word in your variable names, capitalizing the first letter of all words in the name other than the beginning of the variable name. For example, rather than the generic name $average, indicate *what* average the variable holds and use an uppercase letter for subsequent words in the name such as $averageMotifLength.

....

Unlike many other programming languages, Perl does not restrict the type of data that you can store in a particular variable. Said differently, Perl has one simple data type—the scalar. (Two of Perl's more complex data types are the array and hash, both of which we will see later on in chapters 9 and 10, respectively.) Scalar variables can store integers (e.g., 14, −389), real numbers (0.876, 13.456), and strings ("GCGCCG", "Genus and Species"). Although you won't necessarily mix and match different types of data in the same variable, the following lines of Perl are valid and highlight the flexibility that Perl has when dealing with multiple types of simple data.

```
my $someVariable;      # empty at this point

$someVariable = 2;     # first holds a scalar INTEGER
$someVariable = 1.91;  # then holds a scalar REAL
$someVariable = "Pyrococcus furiosus DSM 3638"; # now a string
```

Each variable can only hold one scalar at a time. In the foregoing example, the variable $someVariable is empty at the time of the declaration but then is assigned the integer value two (2). The next statement assigns the real number 1.91 to the variable $someVariable. Because variables can only hold one value at a time, the integer two (2) is *replaced* by the real number 1.91. Once replaced, the integer two (2) is "lost." Finally, a string holding an organism name is assigned into the variable $someVariable, thus the real number 1.91 is replaced by the string.

Box 3.3 Larry Wall's List of Suggestions

It is easy to make up legends about Larry Wall, the inventor of Perl. He does so himself at his Web site, which is an entertaining read and a source of worthwhile links. One of the seemingly plain facts about Larry and his creation is that Perl version 1.0 was introduced to the world on December 18, 1987, a date you may wish to celebrate annually in your lab or classroom. "Hello Perl."

Wall was well integrated in the world of geek creators and was among the many asking questions about why computing had to be the way it was. Perl was developed to deal with text analysis with considerably more finesse than previous attempts in other languages. From the beginning there was a certain renegade freedom in the seminal Perl community. This was a reaction perhaps to the trend of big software makers writing their software in secrecy and charging lots of money for their products. In defiance, Perl can always be downloaded for free, and the Perl community online provides to itself a sort of free and open help desk or IT support network. By the way, Larry's camel book (O'Reilly, 2000) is the Bible of Perl. If you are planning to consider a serious next step in Perl, buy it.

Larry's college years were eclectic. After trying out several different majors (including chemistry and music) as an undergraduate at Seattle Pacific University, he took some time off. When he returned to SPU he designed his own major, "Natural and Artificial Languages." At one point he was planning to be a missionary specializing in Bible translation but was deterred from that career for health reasons.

Larry is in great demand to give keynote talks at conferences. (You can find some of the transcripts online.) A few "pearls" from Larry include one from 1997 at the first Perl conference, Larry's talk was "On the Culture of Perl" (at which he seems to have brought along his own laugh track and other sound effects and used them in strategic places). There he unapologetically (and even triumphantly) outlined his seemingly contradictory points of view on religion and evolution. (He is in complete agreement with evolutionists that we evolved from primordial ooze billions of years ago, and he is a great fan of Daniel Dennet and Stephen Jay Gould.) Do an Internet search for the transcript of his talk and read more if you like. On the topic of linguistics, Larry says he designed Perl as a living, evolving language, which means that like English it is "quirky, sloppy and full of redundancy," and none of those ought to be taken as negative adjectives. "The purpose of language is not to help you learn the language, but to help you learn other things by using the language."

In 1998 at Perl conference 2.0, Larry gave the "2nd State of the Onion." He emphasized the danger of creating too simple and too perfect a language. To summarize the rambling talk that includes lots of graphic images, Perl mimics natural (evolved) languages and that is why it is still evolving and is so versatile. It is why Perl was so readily taken up by a community of programmers and developed at the grass roots.

In short, Larry is a fun Internet search whether to his own site or to the commentaries of others. He is an original thinker and pleasantly outspoken and eccentric as well as just plain brilliant.

Larry Wall has much to say about the influences of natural languages on the development of Perl. The following are his major points in quotations, in the order in which they appear at his Web site. We did a bit of interpretation and annotation, trimmed the list from 16 to 11, and added the numbers to make these look like rules or general principles, which many of them are. This list rambles and flows in Larry Wall style, so feel free to read selectively as needed.

1. "Learn it once, use it many times." That is, Perl was designed to be useful in multiple venues.

2. "Learn as you go." This is good news for Perl novices, such as readers of this book. Just as you learned English a little at a time but began putting it to use almost immediately, so it goes with Perl.

3. "Many acceptable levels of competence." More good news for novices. Larry points out that children or non-native speakers can still get points across in English, even with mispronunciations and grammatical errors. Perl, too, may be written at many levels of competency.

4. "Multiple ways to say the same thing." This *is* a general principle of Perl, and you can depend on it. Nevertheless, beginners may find some comfort in using templates for their first programs.

5. "No shame in borrowing." Great! We expect to do quite a bit of that. Why completely reinvent a chunk of code (or a witty saying) that works nicely as a little module?

6. "Indeterminate dimensionality." Yikes! Larry sometimes lapses into heavy-duty phrasing. It can be summed up as "Natural languages are a kind of fractal and so is Perl!" You might think you are going in a straight line from point A to point B, but the geography in between is going to have a big influence on how you get there. Nevertheless in building a program, point A to B should be one of the first considerations and the fuzzy details in between get refined later. That's a lot of information for new programmers, but it gives us a little glimpse into the thought patterns of a professional.

7. "Local ambiguity is okay." Think about pronouns like *it*. *It* can be ambiguous, but as long as the corresponding noun is nearby, that ambiguity is tolerable. Such local ambiguities get cleared up quickly. A Perl program can be full of local ambiguities, and those too are resolved by local context. In our opinion this is another heavy-duty programming concept. It is nice to know that programmers think this way, but most likely a novice will (and should) exercise caution with ambiguous symbols, variables, and so on.

8. "Topicalization." We were surprised to find that word in the *OED*. Larry did not coin it! It means "to make into a grammatical topic." For example, using a header like what we are doing here by leading this

continued

Box 3.3 Continued

section with the word *topicalization* is in fact topicalization! In an English sentence that might look like "Carrots, I hate them," with *carrots* serving as a topicalizer. So far, so good. However Larry adds that Perl is full of little topicalizers. `foreach` is one of those because it alerts the reader that the following code will be handled in a very particular way. You are free to decide for yourself whether the topicalizer metaphor is useful one for you at this point in time. At the least it suggests that the structure of Perl has some useful hierarchies that may help in organizing a program.

9. "Pronominalization." What did Larry major in at SPU? Oh, right he invented his own major called Natural and Artificial Languages. That is why he has quite a bit more to say on the topic of pronouns (or variables). The flexible structure of Perl is an important theme (see items 7 and 9) and is probably greatly appreciated by professional programmers. As with several of Larry's points, decide where you are with Perl right now before you embrace ambiguity with too much enthusiasm. It might be better to spell things out at first.

10. "No theoretical axes to grind." It doesn't have to be elegant, and (an important message from Larry) redundancy is just fine. Natural languages do not convey information with maximum parsimony and efficiency, and neither does Perl. Natural languages are scattered with redundancy, and this is how points get across. It is the same for Perl. And we would like to add that it is probably the same for DNA. As we try to decipher the meaning in it, we should be ready for lots of inelegance and redundancy.

11. "Style not enforced except by peer pressure." This is a tricky one. Maybe you are a Perl novice and are handing in assignments to a teacher or are teaching yourself and just trying to make sense of it all. Or maybe you are the teacher and students are handing assignments to you. Either way, you might hope for (or even demand) a bit of clarity and conformity in the style just to keep everyone's sanity intact. So that is the peer pressure part. However, Larry points out that Perl is going to work no matter what your personal style is, and you can go ahead and develop your own (if your peers will let you.)

3.4 Built-In Functions

Perl's built-in functions are small Perl programs that have already been written and are included with the language. You might view these built-in functions in much the same way that you view functional buttons provided on a good calculator. For example, a good calculator comes standard with buttons to perform the mathematical operations of square root, logarithm, and y^x. As a user of the calculator, you expect these buttons (built-in functions) to work correctly and efficiently. Perl's built-in functions work in a similar manner. For example, when you use the `length` function, Perl counts the number of characters in the string and returns that number; you as the programmer do not have to write Perl statements to count the number of characters in a string, the built-in function does that work for you. Or if you use Perl's built-in function to calculate a square root, the `sqrt ()` function will do the mathematical work for you.

Perl's built-in functions are collected in a library. Your Perl programs have automatic access to all the built-in functions. Later on, we'll show you how you can write your own Perl functions as well as how to download special-purpose modules of Perl functions written by others.

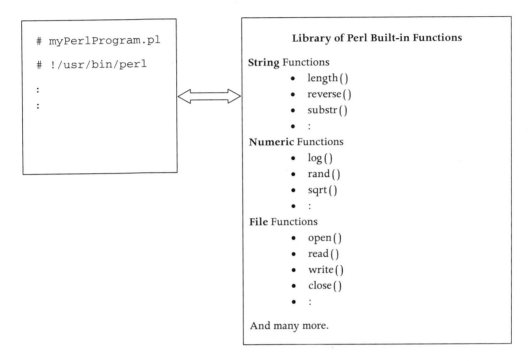

3.5 Strings and Their Built-In Functions

In general, *strings* are groups of characters within double quotes, including spaces. In particular, DNA sequences are strings. As a programming language, Perl has many strengths, but the extensive and flexible uses of and for strings is one of its most powerful features. From individual nucleotides, short and long motifs, and entire genomes, strings are the stuff of DNA sequence analysis, and Perl offers a number of built-in operators and functions to work on strings.

The following examples introduce a set of Perl built-in functions to operate on strings. Although the examples will use small sequences of DNA, any string will work, for example, an English word, sentence, or entire poem. At this introductory point, we will use small sequences of DNA to keep the focus on the operations of the functions and operators, but later we will use these functions and operators on much larger sequences, such as an upstream region of a particular gene or an entire microbial genome.

length(*STRING*)

One of the most common operations performed on a string is to ask for its length. For demonstration purposes in the examples to follow, the string length is explicitly evident in the Perl code; however, we will soon encounter situations where the sequence length is not known until we ask the length function to compute and return the length. Examples of times when a sequence length will be unknown include when your program reads in a motif from a file, snips out an upstream region of sequence, or loads an entire genome.

A detailed explanation of the use of the length function as shown in the next example will serve us well for future understanding and uses of built-in functions. Functions, whether built in or created by you, usually have two distinguishing features: (1) arguments and (2) a return value. The length function for example has one argument, a string. The argument, a Perl string, is the item that is "sent to" the length function when called. A return value is sent back to your Perl program from the function, the length function returning a value indicating the number of characters in the string.

Value returned	Function call
3	length("ATC")
14	length("ACCGTTACGTCGCG")
0	length("")

When your Perl program uses a built-in function, it is referred to as "calling the function." The following Perl program includes five comment lines that tease apart the action that occurs in the Perl statement that makes the call to the function length.

```
$motifLength = length($someMotif);
```

As with all Perl statements, we first address the instructions to the right of the assignment operator (=). In this case, (1) a call is made to the built-in function length. The length function requires one argument, that is, between the parentheses you must provide the length function with a string. In this example, the variable $someMotif holds the string so (2) the string in $someMotif is the argument that is sent to the length function. At this point, your Perl program actually pauses while the length function does its work, in this case (3) counting the number of characters in the string within $someMotif (14). Once the length function has computed the answer, (4) the answer is returned to your Perl program. Steps (1) through (4) occur to the right of the assignment statement. The final step, the returned value of 14, is now the result of the right-hand side. If we could halt your program after returning from the function and before the assignment operator does its work, the state would be as follows.

```
$motifLength = 14 (returned from length)
```

The final step (5) assigns the value returned from length to the variable $motifLength. The print statement would then output the resulting message.

```perl
#!/usr/bin/perl

use strict;
use warnings;

# A program to compute and print the length of a sequence

my $someMotif;

$someMotif = "ACCGTTACGTCGCG";

# (1) call the library function length
# (2) send the string stored in $someMotif to length()
# (3) the length function counts the letters in $someMotif
# (4) the length function returns the answer
# (5) the answer is assigned (=) to the variable $motifLength

my $motifLength = length($someMotif);

print "The motif $someMotif has a length of $motifLength (bp).";
```

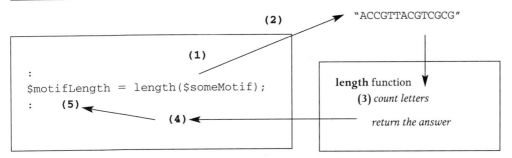

```perl
print "The motif $someMotif has a length of $motifLength (bp).";
```

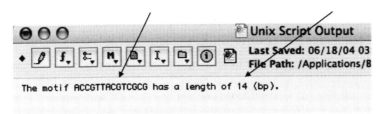

The length function is one of many functions you will encounter. In general, the syntax for using a function that returns a value is shown below.

```perl
$returnValue = functionName(argument list);
```

where the argument list is a list of one or more comma-delimited items that the function needs.

Good Practices

In the previous example, notice that the print function outputs the *values* stored in the variables $someMotif and $motifLength, not their names. The print function "interprets" all the variables within the double quotes and substitutes the number held in the respective variable into the message to be printed. The print function also allows you to enclose the string to be printed in single quotes; however, within single quotes your variables will not be converted to their values. This example is a case where you would *not* want to use single quotes because they do *not* interpret variables within the string and give you literally what you single quote. The difference is shown below.

```perl
#!/usr/bin/perl

use strict;
use warnings;

my $someMotif;

$someMotif = "ACCGTTACGTCGCG";
my $motifLength = length($someMotif);

# variables are interpreted within double-quotes; values printed
print "$someMotif has a length of $motifLength (bp).\n\n";

# variables are NOT interpreted within single-quotes
print '$someMotif has a length of $motifLength (bp).\n\n';
```

The motif ACCGTTACGTCGCG has a length of 14 (bp).

The motif $someMotif has a length of $motifLength (bp).\n\n

index(*STRING, SUBSTRING [, POSITION]*)

The index function will return the location of the first instance of a known target *SUBSTRING* within a larger *STRING*. For example, the index function could help you locate the first adenine (A) nucleotide that occurs in a GC-rich sequence or the first potential ATG codon that codes for the amino acid methionine in a putative coding region. In Perl, all strings have numbered locations, as shown in the following string. Note that the initial character is numbered beginning with zero (0). The index function returns the integer location or negative one (−1) if the target sequence is not found.

"ACCGTTACG"
012345678

Thus, the first cytosine (C) nucleotide is located in position 1, whereas the first occurrence of the dimer CG is in location 2.

The index function has two mandatory arguments and an optional third argument. The first argument is the *STRING* to search and the second argument is the *SUBSTRING* to search for within the (larger) first *STRING*. The optional third argument indicates at what *POSITION* to begin the search. If a third argument is not provided, the search begins at the beginning in position zero (0).

```perl
#!/usr/bin/perl

use strict;
use warnings;

# A program to find the location of the first
# three-letter ATG (start) codon

my $putativeGene; # holds start of a putative coding region

$putativeGene = "ACCGTATGTACG";

my $whereStart; # to hold the location of first start codon

# find the first location of a start codon (ATG)
$whereStart = index($putativeGene, "ATG");

print "First ATG codon found at position $whereStart \n\n";

my $whereAC; # to hold location of first dimer AC after ATG

# after ATG, find the first location of the dimer AC
$whereAC = index($putativeGene, "AC", $whereStart);

print "After ATG, first AC dimer found at position $whereAC \n";
```

Assuming that the variable $whereStart holds the starting position of the initial ATG codon, we can use this knowledge to start another search at this location by using the value in the variable $whereStart as the third argument to the index function. The third argument indicates the position to begin the search. Because $whereStart holds the starting location for ATG and we know AC is not to be found within ATG, we alternately could have started our second index search at ($whereStart+3).

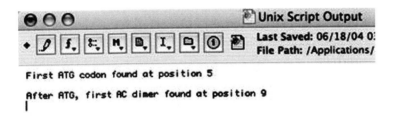

Unix Script Output

Last Saved: 06/18/04 0:
File Path: /Applications/

First ATG codon found at position 5

After ATG, first AC dimer found at position 9

....

Good Practices

Take particular notice of the second call to index. You'll notice that we used the variable $whereStart as the third argument. Why did we do that? Couldn't we just have used the integer 5? Well, in *this* example, an integer 5 would work and the explicit 5 would cause your program to produce the same output. But note that an integer 5 *only* works for the case where we start with the string ACCGTATGTACG. If you or someone else were to ask you to run your program using a *different* sequence where the initial start codon does not start in position 5, then using an explicit 5 as the third argument in the call to index may not produce the correct result. In our solution, we know that the variable $whereStart holds the position of the initial ATG in whatever sequence is held in the variable $putativeGene. If someone changes the sequence in $putativeGene and reruns the program, $whereStart will hold the different ATG starting position. Thus using $whereStart as the third argument to index is a good general solution that works on whatever sequence is used. Perhaps this seems obvious, but from a programming point of view, the distinction between using a specific value (5) and a general solution that uses a value stored in a variable often distinguishes a dangerous program from a good program. Note that we said "dangerous" and not just a "bad" solution. A dangerous program runs and produces answers . . . but the answers are *sometimes* wrong!

....

You may have noticed a potential problem with the given solution. Specifically, the example assumes that the substring ATG *will be found* in the sequence held in $putativeGene. If an ATG is not found, the index function will return a −1. In chapter 6, we'll show you how to write a program that catches the −1 returned from the index function and prints a more appropriate message such as "No ATG was found."

....

Try This Now

1. Run the program that uses the index function.
2. Change the sequence so that the ATG codon appears in a different position. Rerun the program.
3. Change the sequence so that *no* ATG substring appears in $putativeGene. Notice the message that is printed.

....

rindex(*STRING, SUBSTRING* [*,POSITION*])

The rindex function works like the index function except it works in reverse (r), that is, from the right-hand side. rindex returns the last occurrence of target *SUBSTRING* within the larger *STRING*. Because rindex searches for the substring from right to left, if the third argument *POSITION* is included, rindex returns the right-most occurrence of the *SUBSTRING* before or at the *POSITION*.

```
my $putativeGene; # holds start of a putative coding region

$putativeGene = "ACCGTATGTACG";
print "Sequence: $putativeGene \n\n";

my $whereStart; # to hold location of the last cytosine

# find the last location of a cytosine
$whereStart = rindex($putativeGene, "C");

print "The last Cytosine (C) found at position $whereStart";
```

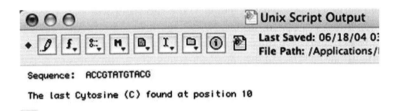

Sequence: ACCGTATGTACG

The last Cytosine (C) found at position 10

lc (*STRING*)

The lc function converts the characters in the *STRING* to lowercase characters and returns the newly converted string. It is often wise to convert all of your DNA sequence to either lowercase or uppercase before proceeding further. Because the lc and uc functions return a newly modified sequence, you can either store the sequence in a new string variable or replace the original string with the newly modified string.

uc (*STRING*)

The uc function converts the characters in the *STRING* to uppercase characters and returns the newly converted string.

```
my $someGene; # holds beginning of a coding region

$someGene = "atgctcgtccgcgcccta";
print "Coding Region (original): $someGene \n";

# (1) convert all the orginial nucleotides to uppercase
# (2) overwrite the original with the new uppercase sequence
$someGene = uc($someGene);

print "Coding Region (after uc): $someGene \n\n";
```

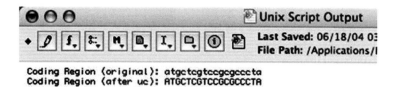

substr(*STRING, OFFSET [, LENGTH]*)

The substring (substr) function extracts a substring from a longer *STRING*. The second argument indicates the starting position or *OFFSET* of the substring to extract. For example, if the second argument was a 2, the substring would begin at the third character in *STRING* (remember that the initial character position is at zero [0]). If the second argument is a negative number, the substring will start that far from the right end of the *STRING*. The optional third argument indicates the *LENGTH* of the substring to extract. If the third argument is omitted, the substr function returns everything from the offset to the end of the string. If the third argument is a negative number, the substr function will leave that many characters off the end of the string.

```perl
my $someGene; # holds beginning of a coding region

$someGene = "atgctcgtccgcgcccta";

# convert all nucleotides to uppercase
$someGene = uc($someGene);

print "Coding Region: $someGene \n\n";

# get the third (position 6) and fourth (position 9) codons
my $codon3;
$codon3 = substr( $someGene, 6, 3 );
my $codon4;
$codon4 = substr( $someGene, 9, 3 );

print "The third codon is $codon3 \n";
print "The fourth codon is $codon4 \n\n";
```

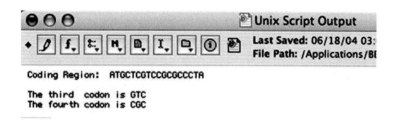

It is common to use index and substr in tandem. For example, index can be used to find the location of the start codon (ATG), and then substr could be used to extract the next two codons as shown below.

```
# find the first location of a start codon (ATG)
$whereStart = index($putativeGene, "ATG");

# snag the 2nd and 3rd codons (just beyond ATG)
$codon2 = substr($whereStart+3, 3);
$codon3 = substr($whereStart+6, 3);
```

reverse(*STRING*)

Reversing a sequence of DNA is a classic requirement for many applications, and not surprisingly this common utility is included as a built-in function in Perl. The reverse function takes one argument, a string, and returns the reverse of that string. The reverse function is versatile enough to handle both a scalar such as a string or a list of items. Although we won't see lists until later on, the reverse function performs as you might imagine on a list, that is, it returns the list of items in reverse order.

```
#!/usr/bin/perl

use strict;
use warnings;

my $someSequence;
my $revSequence;

$someSequence = "ACCGTTACGTCGCG";

print "Original sequence: $someSequence \n\n";

$revSequence = reverse($someSequence);

print "Reversed sequence: $revSequence \n\n";
```

⊖ ◯ ◯   🖹 **Unix Script Output**

• 🖊 *f,* ⌗, M, ▣, I, ◻, ⓘ 🖺   **Last Saved: 06/18/04 03**
                               File Path: /Applications/I

Original sequence: ACCGTTACGTCGCG

Reversed sequence: GCGCTGCATTGCCA

Box 3.4 Why Do Computer Scientists Count from Zero?
....

Let's turn that title question around and ask why the rest of the world is so inconsistent in its counting practices.

Although it makes practical sense that any useful counting system ought to begin with zero, children are persistently taught to begin at one. Perhaps it is the fault of our having such convenient counting objects as our ten fingers and our reluctance to imagine zero of them. Indeed, as books like *Zero* (Seife 2000) and *The Nothing that Is* (Kaplan and Kaplan 1999) attest, the concept of zero was not only not obvious but even strongly resisted through most of the history of human counting.

Nevertheless (index-like) we all begin our lives at age zero and spend our entire first year as age zero, consistent with the way we name subsequent years as age 1 or age 11 or age 51. However a certain discomfort level with zero means that we prefer to avoid it as an infant's age (especially for an entire year) and substitute the (more satisfying) counting of weeks and months instead. The same is true for the Western calendar, which began awkwardly with the year one rather than the year zero, as it should have. Thus there was a bit of scholarly confusion about exactly when we should consider the new millennium to have begun. If the year 1 was really the year 0, then we were premature in opening all that champagne in 2000!

Meanwhile, the practicalities of travel through space lend themselves well to the correct use of zero. We do that instinctively perhaps because the physical reality is so unavoidable. For example, a trip necessarily begins with zero until one has taken the first step. Certainly one does not get to count mile number one or even step number one until one has achieved it. That mundane truth is the simple reason for beginning an index or list or array in computer science with zero. Because any journey begins with zero, it is a simple extrapolation to a trip down the length of a string of As, Cs, Gs, and Ts. Until we have moved off the first DNA base, we are at zero. That starting base, is not our first base; it is our zeroth base!

This elegant (and correct) counting system, affords calculations to be visualized more readily, which is why number lines with zero are eventually (but perhaps not soon enough) introduced in elementary school math. We can slide with ease up and down a number line that commences with zero adding and subtracting numbers (including $1-1 = 0$) and always arriving in the right spot. And unlike a number line beginning with one (if there is such a thing outside of the realm of a child's fingers), zero provides possibilities to explore the realms of negative numbers as well as Zeno's territory of infinite fractions less than one.

3.6 String Operators

In addition to the built-in functions that work on strings, Perl provides a rich set of string operators. Operators work like functions but use different flavors of syntax, for example, the concatenation operator (.) is a binary operator with a syntax similar to mathematical addition (+), whereas the pattern matching operators for substitution (s) and transliteration (tr) use the forward slash character (/) to surround arguments rather than parentheses. Many programmers find these small differences in syntax to be confusing. (including us!) But on a positive note, like the small alterations in syntax and use in natural languages that provide a richness of expression, Perl provides a varied set of functions and operators. The need to refer back to a text or help page for the correct syntactical use is the small price we pay for access to the rich set of operations on strings.

3.6.1 The Concatenation Operator (.)

The period (.) is the binary, infix operator for the concatenation of two strings. It is a *binary* operator because it requires two string operands to work on and *infix* because the operator appears in between the two strings. Note that the concatenation operator does not leave a blank character between the concatenated strings; rather, the last character of the left string immediately precedes the initial character of the right string. As the next example shows, this behavior is desirable when using Perl to generate file names that do not contain blank characters within the name.

```perl
# concatenate genus,species,strain into an Excel (.xls) filename

my $genus = "Escherichia";
my $species = "coli";
my $strain = "K12";

my $nextOrganismFilename;

$nextOrganismFilename = $genus . $species . $strain . ".xls";

print "Filename: $nextOrganismFilename \n";
```

Unix Script Output

Last Saved: 06/18/04 03
File Path: /Applications/E

Filename: EscherichiacoliK12.xls

There are other ways to concatenate strings. Perhaps the most straightforward is to include your string variables within double quotes. Remember that within double quotes, your variables are interpolated, which means the *values* within the variables are used in the resulting string. Note that the period within the double quotes is *not* the string concatenation operator but a literal period needed before the Excel file name extension (.xls). The double quotes alter the context so Perl treats the period as any other character.

```
$nextOrganismFilename = "$genus$species$strain.xls";
```

In addition, Perl's `join` function is an efficient way to concatenate strings. We'll see the `join` function in a later chapter.

3.6.2 `tr`: the Transliterate Pattern Matching Operator

tr/*CHANGETHESE***/***TOTHESE***/**

The **transliterate** (`tr`) operator (referred to here as the translation operator) changes all occurrences of the characters listed in *CHANGETHESE* to the corresponding characters listed in *TOTHESE*. Note that the changes are made in a one-to-one fashion, that is, all occurrences of the first character listed in *CHANGETHESE* are changed to the first character listed in *TOTHESE*. The `tr` operator returns the number of characters that were changed.

```
tr/{/[/         # convert all left braces { to left brackets [
tr/}/]/         # convert all right braces } to right brackts ]
tr/!/./         # convert all exclamation points ! to periods .
tr/ATCG/TAGC/   # convert to the complement sequence
```

You may have noticed that the string on which the transliteration will occur is not an argument in the `tr` operator. Unlike most of the previous functions, the two pattern-matching operators, `tr` and `s`, *bind* to a target string via the two-character binding assignment operator (=~). As shown shortly, the binding operator ~= binds the operator to a sequence and performs the operator on that sequence, for example, translating a sequence to its complement.

```
my $sequence ="ACCGTCTTAG";

print "Original:  $sequence \n";

$sequence =~ tr/ACGT/TGCA/;




print "              |||||||||| \n";

print "Complement:  $sequence \n";
```

The =~ binding operator binds the string $sequence to the pattern in the tr *operator. The* tr *operator performs the transliteration with the altered string residing in $sequence.*

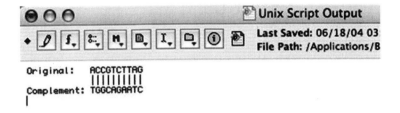

The previous example shows in part how the `tr` operator can be used to quickly transliterate a sequence into its complementary sequence. When used in conjunction with the `reverse` function (see foregoing), a sequence can be converted to its reverse complement. For example, given a sequence of DNA on the direct (+) strand, the `tr` operator followed by the `reverse` function can produce the reverse complement sequence that would be found on the complementary (−) strand.

```perl
# A program to find the reverse complement of a sequence of DNA

my $directStrand; # holds sequence on the direct(+) strand

$directStrand = "ACCGTATGTACG";

print "+ Strand: $directStrand \n\n";

# get complementary sequence
my $comp;
$comp = $directStrand; # make a copy
$comp =~ tr/ACGT/TGCA/; # transliterate

# get reversed complementary sequence
my $complementaryStrand;
$complementaryStrand = reverse($comp);

print "- Strand: $complementaryStrand \n\n";
```

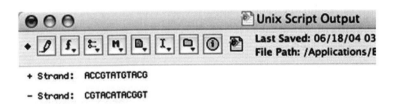

3.6.3 Counting Base Pairs with `tr`: An Advanced Example

An interesting side effect of transliteration with the `tr` operator is that it returns the number of characters that were changed. This can be creatively (and efficiently) used to count the number of characters in a string. To "catch" and save the number of characters that were changed requires an unusual-looking line of Perl.

```
$numC = ( $sequence =~ tr/C/C/ );
```

First of all, when using `tr` to solely count the number of occurrences of a particular character, we don't really want to *change* the character, thus we explicitly translate it to itself. For example, in the example line, the cytosine (C) is changed to itself. The second unusual piece of syntax is the use of both the binding operator (=~) and the assignment operator (=) on the same line. As with all assignments, the instruction is executed from the right to the left, thus first the `tr` operator binds to the string stored in the variable $sequence and then `tr` translates all occurrences of the letter C to a C. On the completion of the `tr` operation (everything inside the parentheses), the number of characters that were translated is returned and then assigned (=) to a variable to hold that count ($numC). Note that the parentheses are optional but help remind the reader that all the work on the right-hand side (inside the parentheses) is done first and *then* the result is assigned to the variable.

```
# count number of pyrimidines

my $sequence = "ACCGCTCTTCAG";

my $numC;
my $numT;

print "\t Number of bases\n";

$numC = ($sequence =~ tr/C/C/);
print "C:\t $numC \n";

$numT = ($sequence =~ tr/T/T/);
print "T:\t $numT \n";
```

```
         Number of bases
C:    5
T:    3
```

a. `tr`'s arguments, *CHANGETHESE* and *TOTHESE*, do *not* do variable interpolation (i.e., when a variable is replaced by the value it holds), thus we cannot use variables in these locations. The arguments to `tr` must be explicitly listed. The example that follows will execute (that is, it is not a Perl syntax error), but the number of nucleotides that get translated will be zero.

```
$nuc = "A";
$numA = ($sequence =~ tr/$nuc/$nuc/); # NO, this will not work
$numA = ($sequence =~ tr/A/A/);       # YES, this will count As
```

b. Although we have not yet covered regular expressions, it is worth noting at this point that regular expressions do not work with the `tr` operator.

....

3.6.4 s: The Substitution Pattern Matching Operator

s/*PATTERN*/*REPLACEMENT*/*[option]*

The substitution (s) operator searches a given string for a particular *PATTERN* and replaces the characters in the *PATTERN* with those characters from the *REPLACEMENT*. This operator returns the number of substitutions made. Of particular significance for this operator is that the *PATTERN* can be a regular expression (see chapter 4), thus this operator provides a rich set of possibilities for crafting patterns that you need to find and replace with alternate characters. Additionally, a number of options are available when performing substitution, including an ability to globally (g) replace all occurrences of the *PATTERN* or to perform case-insensitive (i) pattern matching.

Note that a specific substitution will ultimately occur within larger Perl programs that you will learn to write. For each of the examples that follow, we assume a very specific context to keep the focus on learning how to use the substitution operator.

The most straightforward substitutions are when you know the exact *PATTERN* that you wish to replace. For example, perhaps you'd like to highlight the first putative start codon (ATG) from a sequence by making it appear in uppercase.

....

Good Practices

Because the substitution (s) operator is a "destructive" operator—that is, it permanently alters the string—we will make a copy of the original string prior to the substitution and then perform the substitution on the copy. It is often the case that you'll want the original version later so making a copy and operating on the copy is a wise move.

....

```
my $someGene; # upstream and beginning of a coding region
my $copyGene; # working copy of the region

$someGene = "cgccatataatgctcgtccgcgcccta";

print "Original: $someGene \n\n";

# highlight start codon
$copyGene = $someGene;
$copyGene =~ s/atg/ATG/; # substitute upper for lower case

print "Modified Start Codon: $copyGene \n\n";
```

```
Original:    cgccatataatgctcgtccgcgcccta

Start Codon: cgccatataATGctcgtccgcgcccta
```

To enable a quick view of AT-rich regions, the next example converts adenine (A) and thymine (T) nucleotides to uppercase. The global (g) option is added in this case because we want to *globally* substitute *all* occurrences of lowercase a to an uppercase A.

```
# convert lowercase a and t to uppercase to view AT-rich regions

$copyGene = $someGene;

$copyGene =~ s/a/A/g;  # substitute 'A' for 'a' globally
$copyGene =~ s/t/T/g;  # substitute 'T' for 't' globally

print "AT-rich:    $copyGene \n\n\n";
```

```
AT-rich:    cgccATATAATgcTcgTccgcgcccTA
```

The substitution operator allows regular expressions for *PATTERNs*. After introducing regular expressions in the next chapter (chapter 4), we will return to the substitute operator at the end of that chapter (section 4.10) for more advanced applications of substituting characters.

4 : String Play with Regular Expressions

In which an unconventional first step will be taken. Namely, the reader will be urged over some hurdles to begin using regular expressions immediately and will then find unexpected entertainment and instant gratification with some surprisingly relevant word play.

match (mach), v. To come or bring together as equals or associates. **1.** a. trans. To be equal to; to be equal of in extent or degree; to correspond to; to be the match or fitting counterpart of. Also: to fit exactly or dovetail into. (*Oxford English Dictionary*)

4.1 Plunging In with Regular Expressions

We would argue that almost all DNA sequence analysis is one form or another of pattern matching. (The rest is indexing as we strenuously presented in chapter 1.) Whether you are comparing sequences to build phylogenetic trees or to determine their identities, you are looking for similar patterns. Therefore we take the unconventional step of introducing a powerful syntax for pattern matching immediately. The tools are called *regular expressions* or *regex* for short. Many different programming languages use regexes, however they fit especially well within the syntax of Perl. In the examples of this chapter, the regexes will be embedded within Perl statements. Don't worry that you do not yet know all the Perl in these examples. At this point you will be using verbatim a Perl program that we've already written, and you will just focus on and play with the entertaining parts of these programs, that is, the regexes. Look at almost any other book on Perl and you will find regular expressions in the later chapters because they have been traditionally considered advanced topics. However for DNA sequence analysis, regexes may be just exactly what you hoped for. Therefore, we start here with the instant gratification of a sweet dessert (regex) and move on to more of the meat and potatoes main course in the next chapter.

....

If you have skimmed eagerly ahead to this chapter, failing to stop at chapter 2, go back! You cannot play with our regex examples until you have installed Perl and a Perl programming environment on your computer.

....

The particular datasets that we would like you to play with were generated or captured by us for the purpose of showing off the powers of regular expressions. You will find the data in the form of two long lists at the Web site accompanying this book. One list is a generation of all possible 7-mers (heptamers or seven-letter "words") comprised of As, Cs, Gs, and Ts from the entire genome of the bacterium *Escherichia coli* (*E. coli*). To produce it, we shifted a seven-base (seven-letter) window, one letter at a time, through the entire genome, listing all of the 7-mers, one word per line. We treat the *E. coli* 7-mer list like a word list. It is roughly comparable to the other list you will find at the accompanying Web site: all of the English words in a particular dictionary (*Webster's 2nd International Dictionary* from Mac OS X). Also at the Web site are two Perl programs that you will use in your new Perl programming environment. Instead of writing Perl, you will be writing all sorts of fascinating variations of regexes, embedded, phrase-like, within the Perl syntax. You are essentially filling in the blanks, just as you might if you were using a foreign language phrase book. For example:

"Je voudrais <u>un croissant</u>, s'il vous plait."

You could have all sorts of delicious items inserted for the noun "un croissant": "Je voudrais <u>un café</u>, s'il vous plait." "Je voudrais <u>un brioche</u>, s'il vous plait." You do not need to know much grammar and syntax, yet using the sentences can be quite rewarding and gratifying. In this case, you just ordered breakfast in Paris!

We have taken the foreign language analogy further by using a facing page format, sometimes used to show translations of texts, for example, English translation on the left hand page and French on the right. In this case, the same regexes you try with the list of English words will be tried on the facing page with the list of DNA words, often called motifs.

Box 4.1 Why Repeats Are Interesting

Mirror Repeats (MR)

With your thumbs touching, hold out both your hands as though to grasp a thick rope. That is the model for many DNA binding proteins, the mirrored structure seeking out (or actually happening upon) a corresponding mirror-like sequence in the DNA. Interestingly, the mirror may be in the form of an actual reflecting sequence as in this double stranded sequence

```
ACGT-TGCA
TGCA-ACGT
```

Inverted Repeats (IR)

The reflecting sequences could be more of a funhouse mirror, a twisted helix-like, so that the double stranded sequence (called an inverted repeat) is

```
TAGG-CCTA
ATCC-GGAT
```

Give that mirror sequence (both strands) a little twist at the pivot point and you have a straight mirror!

```
TAGG-GGAT
ATCC-CCTA
```

Meanwhile notice that the properties of the original inverted repeat include the potential for internal base pairing, forming two hairpin-like loops or a sort of cruciform shape. (The rest of the strand is represented with a stretch of As and Ts.)

```
AAAAA TAGG-CCTA AAAAA
TTTTT ATCC-GGAT TTTTT
            GC
            GC
            AT
            TA
      AAAAA   AAAAA
      TTTTT   TTTTT
            AT
            TA
            CG
            CG
```

In some cases the formation of loops by internal base pairing has a regulatory function as in the various bacterial operons for the genes that synthesize amino acids such as tryptophane. In that particular case the looping occurs in the RNA transcript.

Extensive internal base pairing occurs in rRNAs and tRNAs and some of it seems to be functional in introns, which are often full of inverted repeats. Some transposable elements and viruses are flanked by inverted repeats, which may comprise a mechanism for movement.

Direct Repeats (DR)

For a rough approximation of a binding protein of direct repeats, hold your two hands as if to grab the rope again but this time as though you were ready to go hand over hand, the pointer finger of the lower hand touching the pinkie finger of the upper hand. Some direct repeats in DNA may correspond to a direct repeat like symmetry in binding proteins.

```
TAGG-TAGG
ATCC-ATCC
```

Other functions of direct repeats include some transposable elements and viruses that use flanking direct repeats for movement. Furthermore, genetic engineering takes advantage of direct repeats as restriction sites for moving and inserting DNA.

Versatile Repeats (VR)

Some repeats are versatile, being both direct and inverted. For example notice the properties of

```
GTAC-GTAC
CATG-CATG
```

Some versatile repeats may be both mirrors and directs:

```
ATTA-ATTA
TAAT-TAAT
```

continued

Box 4.1 Continued

> Irregular, Imperfect, or Mismatched Repeats
>
> The vast majority of repeats have one or more mismatches. These are computa-
> tionally challenging because of the many decisions that must be made in search-
> ing for them. For example, how many mismatches will be allowed per string
> and if there is more than one way to make a match, which one will be used?
> These are some of the same issues encountered in any string comparison for
> analyzing phylogenies and relatedness.

4.2 How Comparable Are These Two Word Lists?

A list of words from an English dictionary and a list of seven-letter DNA motifs from *E.
coli* are not directly comparable at all. However, they are useful as introductory datasets
for learning about regex syntax and thinking about how pattern matching might be
applied to DNA sequences. The beginning of each of the two lists are shown next.

The list of DNA motifs (words) was generated artificially from all possible 7-mers in
the *E.coli* genome. It is a list with little meaning in and of itself, a collection in which all
seven-letter motifs that appear in the *E.coli* genome are considered. That is not to say that
none of these 7-mer motifs are without meaning. A few DNA motifs (of this length) have
had their functions worked out and demonstrated in the lab. For example, some are pro-
tein-binding sites and some are restriction sites; however, the majority of short motifs
remain to be analyzed, and generating motif lists from a genome is merely one small step.

The list of English language words is full of meaning to any native speaker. It
serves as a sort of familiar, comfortable grounding for your first forays into pattern
matching. However it is not meant to be directly analogous to the DNA motif list.
Any English words that you "discover" will already have well-established meanings.
In fact, in this chapter you are not making much of a discovery at all but are testing
your ability to write an accurate search statement using the regex syntax. Yet this is
quite important because regexes carelessly written can return bad results rather
quickly, and it takes some practice and vigilance to spot the errors. In this case, we
use our knowledge of English to spot the successful searches and the errors (if any).

```
english.txt
A
a
aardvark
aardwolf
Aaron
Aaronic
abacate
abacay
abacinate
abacination
abaciscus
:
:
```

ecoli.fna

AAAAAAA

AAAAAAC

AAAAAAG

AAAAAAT

AAAAACA

AAAAACC

AAAAACG

AAAAACT

AAAAAGA

AAAAAGC

AAAAAGG

:

:

4.3 What Would Be a More Comparable Pairing of Lists?

If we did not know any English or if we had taken up the task of deciphering an unknown language such as the ancient and cryptic Linear B, we might take a first step of choosing a text and generating from it all possible words of certain lengths. Then we would examine the frequencies with which words and word patterns occurred. Certain useful everyday words might be expected to occur often. Certain elements of grammar and syntax might be expected in repeating patterns. Some aspects of the analogy might be useful for DNA sequence analysis, an idea we return to in later chapters.

Box 4.2 Alice Kober, Linear B, and Motif Frequencies

....

Linear B is a writing system of 90 characters for an ancient Greek dialect in use about 3500 years ago. It was discovered written on clay tablets in 1900 and remained undeciphered until 1952. The laborious primary method for decoding the script was that of Alice Kober (1906–1950). She worked entirely without a computer, of course, but she was further encumbered by the World War II rationing of paper. Kober created letter and word frequency lists by hand on more than 180,000 notecards, cut to a size of two by three inches from scraps of paper. These scraps included envelopes, greeting cards, and exam books. The relevance of the size was that the cards fit into empty cartons from Lucky Strike and Fleetwood cigarettes, smoked by Kober. Her death at age 43, probably from cigarettes, predated by two years the announcement of the decoding of Linear B. She received little or no posthumous credit at that time. A colleague, Michael Ventris, was awarded an Order of the British Empire for the work. However the method used for deciphering Linear B is sometimes called the Kober/Ventris approach and is potentially relevant to other strings of text, including DNA. For example, generating a motif frequency list of all possible motifs (within a certain size range) is now a relatively simple first step, with the help of computers.

Source: Based on Palaima and Trombley, 2003.

4.4 Introducing the Syntax

Using regular expressions for the exploratory analysis of texts involves two steps:

1. Define a pattern to match in a text using regular expression syntax, for example,
   a. to find all words ending in "omics."
   b. to find all motifs that are GC-rich.
   c. to find all files in a directory that end with the extension . fna.
2. Apply that regular expression to the text in question, for example, to each of the words in a dictionary, poem, or story or to all putative regulatory motifs upstream of a particular gene.

This chapter focuses on the first step, the syntax of regular expressions that you can use to define your pattern.

The following is an example featuring a rich set of regular expression syntax. Do not feel daunted by the syntax at this point; rather, try to get a sense of the power of finding words that contain certain patterns. Assuming we are searching through a file of English words (e.g., a dictionary), the regex below matches all words in the file that

"start with 'ge' and end with either 'ne' or 'me'."

`^ge.*[nm]e$`

More explicitly, this pattern matches any word that starts (^) with the letters 'ge', followed by any number of letters (.*), and the word ends ($) with either an 'n' or 'm' ([nm]) followed by the letter 'e'. This pattern would match with many words, a partial list of which is shown here:

> :
> :
>
> gelatine
> gene
> genome
> genuine
> germane
>
> :
> :

Again, do not feel intimated by the syntax at this point. The point of this entire chapter is to introduce the pattern-matching syntax of regular expressions in a step-by-step fashion with lots of opportunities for you to practice.

4.5 Setting Up a Perl Program to Let You Play with Regex

The following URL provides directions on installing Perl programs to facilitate and guide your focused practice with regex syntax. In addition to the two Perl programs, this download will also provide a dictionary of English words and a list of 7-mers found in the *E.coli* genome.

1. If you have not installed Perl and a Perl programming environment on your computer yet, you should return to chapter 2.
2. Visit this URL:

   www.oup.com/us/perlforDNA

3. Click on the Playing with regex link (see diagram below).

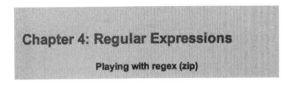

4. Save the downloaded zipped file (regexPlay.zip) on your Desktop.

5. Note: If the file zipped file does not automatically open,
   a. Double-click the zipped file.
   b. Create a New Folder (regexPlay) on the Desktop, and then
   c. Extract the files.
6. You should see a new directory, regexPlay, on your Desktop. Open the directory. You should see the following four files:

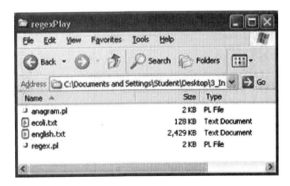

regex.pl	Perl program to use regex to find patterns (use first)
anagram.pl	Perl program to use regex to find anagrams (used later)
english.txt	English dictionary (list of words one per line)
ecoli.txt	7-mers found in E.coli genome

7. Open your Perl programming environment.
8. From the File menu, select Open and open the Perl program: regex.pl. This Perl program should appear in your programming environment.

9. Each time you would like to try a new regex or use a different dictionary, you will only need to modify the following lines. To use a new regex, change the pattern between the single quotes. Note: You *must* keep the single quotes! By the way, this particular regex means "match all words that contain a 'q' but are not immediately followed by the letter 'u'."

```
my $regex     = 'q[^u]';
my $filename  = "english.txt";
# my $filename = "ecoli.txt";
```

You must also choose the file to search, either the English dictionary (english.txt) or the list of 7-mers in *E. coli* (ecoli.txt). You should put a comment (#) symbol in front of the line that you do *not* want to use. In the example above, we are searching the file english.txt, thus we have inserted a # to "comment out" (shut off) the ecoli.txt file.

....

In Mac OS X, if you are using the TextWrangler editor, after you change the regex expression and before you rerun the Perl program, you must save your changes to the file.

....

10. Once you have entered the regex and selected the file to search, run the Perl program and view your results. A sample is shown below.

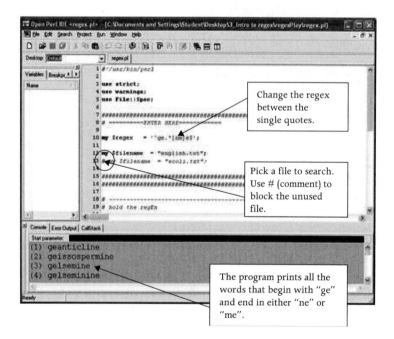

11. If you wanted to search the list of motifs in ecoli.txt, you might change the regex (to be DNA-specific), remove the comment in front of the ecoli.txt line, and insert a comment (#) at the beginning of the english.txt line, as shown here.

```
my $regex = '^GC.*T$';

# my $filename = "english.txt";
my $filename = "ecoli.txt";
```

12. If you make a syntax error (e.g., a typing mistake, such as you forget to include a closing single quote after the regex), the Perl program will complain. If you do not see the output window print results, you can be pretty certain that you have made some small mistake. Your Perl programming environment will typically highlight a mistake that you make in your code. For example, in the program below, I have changed the regex but mistakenly removed the ending single quote and I forgot to reenter the ending single quote. When I run the program, the Perl interpreter recognizes that something is amiss and issues an error message (Can't find string terminator "'" anywhere before EOF at regex.pl line 32) and tries to highlight the incorrect line.

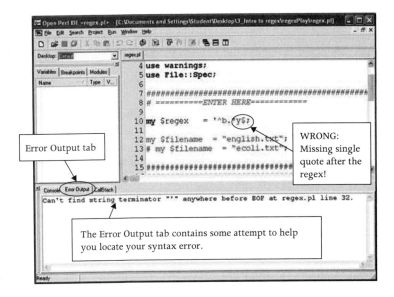

Error Output tab

WRONG: Missing single quote after the regex!

The Error Output tab contains some attempt to help you locate your syntax error.

4.6 Basic Rules of Regular Expression Behavior

Before we begin, a simple description of the basic rules of regular expressions is in order. As the chapter proceeds, we gently introduce you to the various types of regex behavior and encourage hands-on practice with each new set of features.

1. Most characters match themselves, for example, a G matches a G, a T matches a T.
2. A match anywhere within a string is a match. That is, unless explicitly requested to do so, a regex need not match an entire word or line.
3. Anchors (^ and $) can restrain where a match takes place, for example, $ means at the end of a word.
4. Some symbols (e.g., ^) have multiple meanings depending on the context of their use within the regex.
5. Quantifiers (such as + and *) modify the previous character in the regular expression.
6. The match is case sensitive, that is, lowercase c does not match uppercase C. Note, however, that our program regex.pl in this chapter is case insensitive.

4.7 Summary of Regular Expression Syntax

This is a good page for a bookmark—or photocopy the page, laminate it, and use it as the bookmark! Rest assured that we will slowly introduce these features. Note that a more elaborate version of this table appears in chapter 12.

Regex	Meaning
TATA	match four consecutive letters, TATA
TAG|TGA|TAA	match TAG or TGA or TAA
.	match any character but not a newline character
..	match any two characters (independently, not necessarily the same character)
(.)	capture (remember) and match any character
.*	match any character 0 or more times (each is independent of others)
(.*)	capture and match any character 0 or more times
.+	match any character 1 or more times (each is independent of others)
(.+)	capture and match any character 1 or more times
\1	recall the first captured (parenthesized) group
\2	recall the second captured group
\n	recall the *n*th captured group

Regex	Meaning
`.?`	optional, match any character 0 or 1 time
`T?`	optional, match a T or nothing
`(CAAT)?`	optional, match CAAT or nothing
`A{3,7}`	match between 3 and 7 As
`A{3,}`	match of 3 or more As
`[CG]`	match any *one* of the characters in the set, a C or a G
`TATA[AT]`	match TATA followed by an A or a T
`[^CG]`	match any *one* character that is *not* in the set, not a C and not a G
`[CG]{5,10}`	match a C or a G between 5 and 10 times
`^ATG`	string begins with ATG
`TAG$`	string ends with TAG
`\s`	match any whitespace character (tab, space, newline)
`\S`	match any character that is not whitespace
`\d`	match any character that is a digit, same as `[0123456789]`
`\D`	match any character that is not a digit
`\w`	match any one "word" character (includes alphanumeric, plus '_')
`\W`	match any one nonword character

4.8 Facing Page Translations

Facing page translations are often used to help readers translate between two different languages, for example, English text on the left-hand, even-numbered pages and the comparable French text on the right-hand odd-numbered pages. Here we introduce regular expression syntax in small doses, first applying regex to English text on the left and then applying that same regex to DNA sequences on the right. Using regex to find patterns in English words is instantly gratifying. Applying the same regex to find patterns in motifs complements the power of such simple pattern-matching syntax. Each pair of facing pages introducing new regex syntax with examples and questions is followed by an associated pair of pages with answers to the questions.

Scoring in Scrabble

Question: **Are there any words in an English dictionary where the letter q is not immediately followed by the letter u?**

Pattern to match expressed in English: Search for two adjacent characters: the letter q immediately followed by any character that is not the letter u.

[ ]	match any *one* of the characters in the set
[^]	match any *one* character that is *not* in the set

```
q[^u]
 Iraqi
 Iraqis
 Qatar
```

Note: In these facing page examples, our Perl program is instructing each regex to ignore the case of letters, thus you need not worry about upper-versus lowercase letters when crafting your regex. Thus, in the example above, q is really handling [Qq], both upper and lowercase q.

Note: Although these examples are keeping the focus on the regex (rightly so!), remember that you are embedding each regex into a larger Perl program (regex.pl). The Perl program is doing a lot of work for you. Mainly it applies the regex that appears in the single quotes **q[^u]** to *each* word in the dictionary file and prints any of those words that contain a two letter substring where the letter q is followed by a character other than the letter u.

....

Going Back for More

a. All words that contain the three-letter string "ghi".

> *Note:* Do not confuse this with the request to find words that find any one of the letters g, h, or i. In this example, [ghi] is *not* the regex that you want because [ghi] means find words with one of the letters in this set.

> *Note:* If you run a Perl program and it runs on and on, you'll need to halt the program. Depending on your operating system and environment, the method to halt a "run away" program is different:

> **Mac OS X**: Select the keys: Option-Apple-Escape, select the application, and click Force Quit.

> **Windows** (Open Perl): From the Run menu, select **Terminate**.

b. All words with yz that are not immediately followed by an e or an i.
c. All words that contain four consecutive vowels.
d. All words... (*Note:* We strongly encourage you to create your own questions and answers. We are encouraging you to generate regex solutions to your own queries).

....

Scrabble in DNA Land

Our context for DNA Land is a file with one DNA motif (or word) per line. The file is comprised of 7-mers, each line holding one DNA sequence or motif of length seven (7), where a motif is defined as "a pattern with putative biological meaning." The 7-mers were collected by us from NCBI's publicly available DNA sequence for the *E. coli* bacteria. But of course, later on this is where you could substitute your favorite collection of motifs.

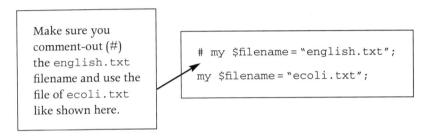

Make sure you comment-out (#) the `english.txt` filename and use the file of `ecoli.txt` like shown here.

```
# my $filename = "english.txt";
my $filename = "ecoli.txt";
```

Question: **Are there any 7-mers in *E. coli* where the nucleotide G (guanine) is not immediately followed by the nucleotide T (thymine)?**

Search for two adjacent nucleotides: guanine G followed by any nucleotide that is not thymine T.

G[^T]
TAAAGAA
ACGTGCC
GATATTT
:
:

All Motifs	**G[^T]**
TCAGTGT	no
GTTCACG	no
TAAA**GA**A	**yes**
AGTAGTG	no
CTTTTTT	no
ACGT**GC**C	**yes**
ACTCATT	no
:	
:	

....

Going Back For More
a. All 7-mers that contain **GTGAC**.
b. All 7-mers that contain the dimer **GC** and this dimer is not immediately followed by a **G** or a **C**.
c. All 7-mers where...

....

Answers to Scoring in Scrabble (English Word Play)

Going Back For More

a. All words that contain **ghi**.

   ghi
   ```
   coughing
   laughing
   laughingly
   laughingstock
   outweighing
   sighing
   weighing
   weighings
   ```

b. All words with yz that are not immediately followed by an e or an i.

   yz[^ei]
   ```
   analyzable
   Byzantine
   Byzantinize
   Byzantinizes
   Byzantium
   unanalyzable
   ```

c. All words that contain four consecutive vowels.

   [aeiou][aeiou][aeiou][aeiou]
   ```
   aqueous
     :
   exsanguious
     :
   Hawaiian
     :
   igneoaqueous
     :
   onomatopoeia
     :
   ```

Answers to Scrabble in DNA Land

....

Going Back for More

a. All 7-mers that contain GTGAC.

GTGAC
TGTGACG
GTGACGT
GTGACGG
AGTGACC
GTGACCA
GTGACAG
:
:

b. All 7-mers that contain the dimer **GC** and this dimer is not immediately followed by a **G** or a **C**.

GC [^GC]
GCTCAAT
TTGCTCA
TGGCACA
AAGGGCT
TGAAGCT

....

Direct Repeats

Question: **Are there any words in which a sequence of three letters is repeated elsewhere in the word?**

Search for words where a sequence of three letters is repeated. This pattern might appear in the middle of a longer word.

```
(.)(.)(.).*\1\2\3
acclimatization
agglutinating
aggressiveness
Albuquerque
alfalfa
allegorically
amalgamate
amalgamated
:
:
```

.	Match any character except a new line
(.)	Match any character and remember it
.*	Match any character zero or more times
.+	Match any character one or more times
.?	Match any character zero or one time
\1	Recall the character from the first parenthesized match
\2	Recall the character from the second parenthesized match
\3	Recall the character from the third parenthesized match

....

Going Back for More

a. Are there any words in which a sequence of two letters is directly repeated at least three times in the word?

b. Are there any words . . .

....

Direct Repeats in DNA Land

Question: Are there any 7-mers in the DNA sequence of *E. coli* where a sequence of three nucleotides is repeated elsewhere in the motif?
Search for motifs where a sequence of three nucleotides is repeated.

```
(.)(.)(.).*\1\2\3
```
AGTAGTG
CTTTTTT
GAAAGAA
GAACAAC
AAAAAAA
GAAGAAA

....

Going Back for More

a. Are there any motifs in which a sequence of two nucleotides is repeated at least three times in the motif?

b. Are there any motifs . . .

....

Answers to Direct Repeats (English Word Play)

Going Back for More

a. Are there any words in which a sequence of two letters is repeated at least three times in the word?

```
(.)(.).*\1\2.*\1\2
anticompetitive
antidisestablishmentarianism
confrontation
confrontations
contentment
enlightenment
inclining
indoctrinating
infringing
insinuating
intercontinental
```

Answers to Direct Repeats in DNA Land

Going Back for More

a. Are there any motifs in which a sequence of two nucleotides is repeated at least three times in the motif?

```
( . ) ( . ) . * \ 1 \ 2 . * \ 1 \ 2
CTTTTTT
TCTCTCG
AAAAAAA
TTTTATT
TTTTCTT
CAAAAAA
AAAAAAA
TTTTTTT
ATTTTTT
ATTTTTT
TAAAAAA
TGAGAGA
TTTTTTG
TTTTTTT
TAAAAAA
AAAAAAT
AAAAAAA
AAAAAAA
TTGTTTT
AAAAAAA
TATATTA
AATATAT
  :
```

Mirror Repeats (Also Called Palindromes in English)

Question: Are there any words in which the first three letters of the beginning of a word are the same as the reverse of those letters at the end of the word?

Search for words where the initial sequence of three letters ends with the reverse of those letters.

```
^(.)(.)(.).*\3\2\1$
despised
detected
deteriorated
detested
foolproof
redder
reviver
revolver
rotator
```

`^ab`	Match ab at the beginning of a word
`xyz$`	Match xyz at the end of a word
`^abs.*[aeiou]$`	Match words that begin with abs and end with a vowel
`\2`	Recall the character from the second parenthesized match

Note: A regex is typically matched against an entire line of text or a long sequence of DNA, not just a single word or motif like these examples. Use of the anchors ^ and $ mean match at the beginning or end of the *entire* sequence, respectively.

....

Going Back for More

a. Are there any words in which three consecutive letters anywhere in a word are followed by the reverse of those letters anywhere in the word? *(Hint: You do not need ^ [start of word] and $ [end of word] for this.)*

   (Note: The following two examples combine both direct and mirror patterns.)

b. Are there any words in which a pair of letters is followed by a direct repeat of those pair of letters followed by two occurrences of the reverse of the pair of letters? (Each of the pairs can be zero or more letters apart, e.g., "AB...AB... BA...BA.")

c. Are there any words in which this pattern occurs twice: a pair of letters is followed by its reverse of those pair of letters? (Each of the pairs can be zero or more letters apart.)

....

Mirror Repeats in DNA Land

Question: Are there any motifs in which the first three nucleotides of the beginning of a motif are the same as the reverse of those letters at the end of the motif?

Search for motifs where the reverse of the initial three nucleotides is repeated at the end of the motif.

```
^(.)(.)(.).*\3\2\1$
AAAAAAA
ATGTGTA
AAAAAAA
TTTTTTT
AAGTGAA
GAAAAAG
TTTTTTT
TTTATTT
TTTATTT
:
```

^ATG	Match ATG at the beginning of a motif
TAC$	Match TAC at the end of a motif
\2	Recall the character from the second parenthesized match

....

Going Back for More

a. Are there any motifs in which three consecutive nucleotides anywhere in a motif are followed by the reverse of those nucleotides anywhere in the motif?
   (*Hint: You do not need* ^ [*start of motif*] *and* $ [*end of motif*] *for this*).
   (*Note: The following two examples combine both direct and mirror patterns.*)

b. Are there any motifs in which a pair of adjacent nucleotides is followed by a direct repeat of those pair of nucleotides followed by two occurrences of the reverse of the pair of nucleotides? (Each of the pairs can be zero or more nucleotides apart, e.g., GT...GT...TG...TG.)

c. Are there any motifs in which this pattern occurs twice: a pair of adjacent nucleotides is followed by its reverse of those pair of nucleotides? (Each of the pairs can be zero or more nucleotides apart.)

....

Answers to Mirror Repeats (Also Called Palindromes in English) (English Word Play)

....

Going Back for More

a. Are there any words in which three consecutive letters anywhere in a word are followed by the reverse of those letters anywhere in the word? (*Hint: You do not need ^ [start of word] and $ [end of word] for this*).

```
(.)(.)(.).*\3\2\1
addresser
addressers
amalgamate
:
analyticity
assertiveness
assesses
Belleville
Brenner
:
```

b. Are there any words in which a pair of letters is followed by a direct repeat of those pair of letters followed by two occurrences of the reverse of the pair of letters? (Each of the pairs can be zero or more letters apart, e.g., AB...AB...BA...BA.)

```
(.)(.).*\1\2.*\2\1.*\2\1
senselessness
```

c. Are there any words in which this pattern occurs twice: a pair of letters is followed by its reverse of those pair of letters? (Each of the pairs can be zero or more letters apart.)

```
(.)(.).*\2\1.*\1\2.*\2\1
Kinnickinnic
```

....

Note: Remember that the character class syntax **[ ]** means to match only one character in the set and **[^]** means to match only one character that is *not* in the set. However, the caret ^ at the start of a regex means match at the beginning of the string. Consider the following (understandable) misunderstanding of using character classes and start anchor.

Find all strings that do *not* contain a traditional vowel: a, e, i, o, or u.

`[^aeiou]`	Wrong! This will match any strings that contain at least one character that is not a vowel.
`^[^aeiou]$`	Probably wrong! This only matches strings of length one.
`^[^aeiou]+$`	Right. All (one or more) characters are not a vowel. Try it!

Answers to Mirror Repeats in DNA Land

．．．

Going Back for More

a. Are there any motifs in which three consecutive nucleotides anywhere in a motif
    are followed by the reverse of those nucleotides anywhere in the motif?
    (*Hint: You do not need* ^ [*start of motif*] *and* $ [*end of motif*] *for this.*)

    ```
    (.)(.)(.).*\3\2\1
    CTTTTTT
    AAAAAAA
    ATGTGTA
    GTGAAGT
    CAAAAAA
    ATAATAA
    AAAAAAA
    TTTTTTT
    ATTTTTT
    AAGTGAA
    ```

    (*Note: The following two examples combine both direct and mirror patterns*)

b. Are there any motifs in which a pair of adjacent nucleotides is followed by a direct
    repeat of those pair of nucleotides followed by two occurrences of the reverse of
    the pair of nucleotides? (Each of the pairs can be zero or more nucleotides apart,
    e.g., GT...GT...TG...TG.)

    ```
    (.)(.).*\1\2.*\2\1.*\2\1
    ```

    *There are none because this regex matches a motif of minimum size of eight, but all
    our motifs are 7-mers! If we had had 8-mers on the list, a match could have been*
    ACACCACA.

c. Are there any motifs in which this pattern occurs twice: a pair of adjacent
    nucleotides is followed by its reverse of those pair of nucleotides? (Each of the
    pairs can be zero or more nucleotides apart.)

    ```
    (.)(.).*\2\1.*\1\2.*\2\1
    ```

    *There are none because this regex matches a motif of minimum size of eight, but all
    our motifs are 7-mers!*

．．．

4.9 More Fun with Regex: Anagrams

> *anagram, n.* 1. A transposition of the letters of a word, name, or phrase, whereby a new word or phrase is formed.

We admit it: we couldn't resist including anagrams. We love word play. In the spirit of this chapter, finding anagrams is a creative and fun application of pattern matching, not to mention an advanced application of some of the regexes you have been learning so far. No worries if you do not understand all the small details of this use of regex. Our reason for including it here is to encourage your creative application of regex for finding patterns in text or DNA sequence.

You need to use the anagram.pl program for this exercise (see section 4.5 for directions to download our Perl program).

The Perl program anagram.pl finds partial anagrams, meaning that not every letter of a word needs to be used to count for an answer, for example, "go" is a partial anagram of "genome."

One-word partial anagrams as defined here require two (2) regexes. $patternOne is a regex to match words as we have been doing in previous examples while $patternTwo is an *inverse* match, that is, keep all words that do *not* match this pattern.

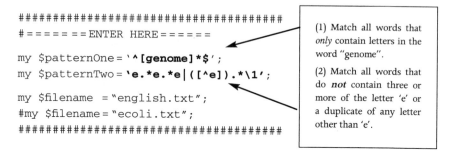

```
#####################################
# = = = = = = ENTER HERE = = = = = =

my $patternOne = '^[genome]*$';

my $patternTwo = 'e.*e.*e|([^e]).*\1';

my $filename  = "english.txt";
#my $filename = "ecoli.txt";
#####################################
```

(1) Match all words that *only* contain letters in the word "genome".

(2) Match all words that do **not** contain three or more of the letter 'e' or a duplicate of any letter other than 'e'.

What Other Words Can You Spell Using the Letters in the Word *Genome*?

Find partial one-word anagrams from the word *genome*.

.*	Match zero or more of *one (1)* character.
[...]	Match *one (1)* character from the set of letters
[...]*	Zero or more matches from the characters from the set of letters *or* the set of strings that contain *only* these letters
^ab	Match 'ab' at the beginning of a word
ing$	Match 'ing' at the end of a word

We can find partial one-word anagrams in two steps (Aho and Ullman, 1995). The result from $patternOne can be the starting list of words for $patternTwo.

$patternOne = *Collect all the words in the dictionary that only use the letters from the set [genome] (Note: You really do need ^ and $).*

^[genome]*$

```
:
egg
ego
:
gene
gnome
:
noon
:
one
```

From this partial list of answers using the $patternOne regex, you can see that some words are listed that should not be included in a final solution, in particular, the letter g appears twice in "egg" (but only once in "genome") and the letters n and o appear twice each in "noon" (but n and o appear only once in "genome"). On the other hand, "gene" is a valid answer because the letter e does appear twice in "genome." This is the reason we need two regular expressions in our solution. The first regex will select only those words that are comprised of letters in the original word, and the second regex will filter out any of those words that use too many repeats of the letters that are not repeated in the original word.

The second regex will assume a starting subset of words that only contain letters from the set **[genome]**.

$patternTwo = *Remove those words with more than two 'e's or with a repeat letter.*

e.*e.*e|([^e]).*\1

Remember that the second regex is an inverse pattern, that is, the Perl program anagram.pl will *not* keep words that match this (inverse) pattern.

---

anagram.pl "joins" two regex together by applying the first and then the second. The resulting output from the first regex becomes the starting list of words for the second regex. The curious reader may want to study the Perl code in anagram.pl at this point. In later chapters, you will learn to fully understand this Perl example.

```
$patternOne    =    '^[genome]*$';
$patternTwo    =    'e.*e.*e|([^e]).*\1';
      :
    ego
      :
    gene
    gnome
      :
    one
```

Going Back for More

Two anagrams that we like are: (listen—silent) and (astronomer—moonstarer).

   a. What words can you create from the letters in your first name? Your full name?
   b. What partial anagrams that form English words can you create from the nucleotides [ACGT]?
   c. What English words can you create from the nucleotides [ACGT] and [N]? N is the symbol typically used to mean "unknown nucleotide," that is, at the time that this genome was sequenced, an N indicates "some nucleotide not determined at this time."
   d. Using a motif from the *E. coli* genome, for example, GCGGC, find motifs that are GC-rich and with a similar number of Gs and Cs.
   e. Augarde's *Oxford Guide to Word Games* (2003) considers anagrams that use multiple words, for example from "genome" we can concoct "me go" and "no Meg." An *appropriate* anagram "should be an appropriate description of the person or thing being anagrammed" (p. 76), for example, from "sweetheart" we get "there we sat," and from "poetry" we get "try Poe" (Agee, 2000). *Antigrams* are anagrams that create phrases that are the opposite of the anagrammed word, for example, from "astronomers" we get "no more stars" and from "funeral" we get "real fun." Consider the additional complexities beyond pattern matching of getting a computer to find an appropriate anagram or an antigram? How might this be analogous to determining the biological function of a certain DNA sequence or motif?

4.10 Using Regex with the Substitution Pattern Matching Operator

s/REGEX/REPLACEMENT/[option]

As introduced in chapter 3 (section 3.6), the **substitution** (s) operator searches a given string for a particular *PATTERN* and replaces the characters in the *PATTERN* with those characters from the *REPLACEMENT*. Often you'll need to find a pattern in a string, replace the pattern, and then throw away all other text surrounding the original pattern. Because the substitution operator (s) allows regular expressions for *PATTERN*s, we can introduce a small amount of regex in our *PATTERN* to help us.

For example, suppose we are scanning summaries of genomes that have been sequenced and are creating a list of the names of the sequencing centers where each genome was sequenced. Some of the sequencing centers names are abbreviated (e.g., TIGR), however, we would like to list the complete name (e.g., The Institute for Genomic Research). In this case, we can search for all occurrences of the string TIGR, replace it with the full name, and remove all other surrounding text. Notice how we ignore all the text before and after TIGR by using the regex syntax to match "any number of characters" (.*).

```
my $summary = "AE017334 NC_00753 05/20/2004TIGR T P C";

print "Long summary: $summary \n\n";

$summary =~ s/.*TIGR.*/The Institute for Genomic Research/;

print "Sequencing Center: $summary \n\n\n";
```

Long summary: AE017334 NC_007530   05/20/2004 TIGR   T   P   C

Sequencing Center: The Institute for Genomic Research

The next set of examples convert the full genus name for an organism (e.g., *Escherichia coli* K12) to the abbreviated genus name first letter followed by a period (e.g., *E. coli* K12). The first example assumes *Escherichia coli* specifically, and the latter examples use regex to handle the general case of using the first letter in the genus name no matter what the name.

```
my $fullName = "Escherichia coli K12";
my $shortName = $fullName;

$shortName =~ s/Escherichia/E./;

print "Fullname: $fullName \n";
print "Short name: $shortName \n";
```

Fullname: Escherichia coli K12
Short name: E. coli K12

Using a regex, we can produce a more powerful and general solution to this problem of using only the first letter of the genus name as shown in the next example. The regex in the *PATTERN*

```
s/^(.)\S+(.*)/$1.$2/
```

does the following: starting at the first letter (^), "save" or capture any initial character (.), ignore one or more (+) of the remaining characters in this name that are *not* whitespace (\S), and then capture the remaining characters in the string (.*). (Note that "match characters that are not whitespace" in regex is represented as *backslash uppercase* S; *we would use lowercase* \s *to mean "match any whitespace".*) The *REPLACEMENT* string, /$1.$2/, begins by recalling the first captured pattern ($1), in this case that pattern is the initial letter (.) of the genus name. This is followed by the insertion of a literal period, followed by the second captured pattern ($2), which is the remainder of the name (.*) beyond the genus name.

The reader should take special care to note that the recall variables in the *REPLACEMENT* section of the substitution operator must be referenced $1, $2, and so on and not \1, \2. Remember that \1 and \2 variables are the way to recall previous matches *within* the same regular expression (see the examples for direct repeats in section 4.8). To recall a captured match, for example, (.) or (.*), after the regex has finished, we must use the Perl variables $1, $2, and so on to recall the respective match.

In the next example, two different organism names are used to show that the identical regex handles both cases.

```
$shortName = "Escherichia coli K12";

# keep only the first letter of the Genus name
$shortName =~ s/^(.)\S+(.*)/$1.$2/;

print "Fullname: $fullName \n";
print "Short name: $shortName \n\n";

$fullName = "Oryza sativa"; # rice
$shortName = $fullName;

$shortName =~ s/^(.)\S+(.*)/$1.$2/; # same RegEx

print "Fullname: $fullName \n";
print "Short name: $shortName \n\n";
```

```
Fullname:    Escherichia coli K12
Short name:  E. coli K12

Fullname:    Oryza sativa
Short name:  O. sativa
```

Notice that $2 captures leading whitespace, that is, all characters after the genus name. For example, $2 holds a single blank character before "coli K12." So $2 ensures that we will get a space between "E." and the final part of the name when we print the results in $shortName.

4.11 Recalling Captured Matches after the Regex

In addition to recalling regex matches within the substitute operator as shown, it is important to appreciate that Perl's $1, $2,...$n variables retain their captured values from the most recent regex match until the next regex. This means that you can recall in subsequent Perl instructions any part of the last regex match that was captured. Because Perl remembers your captured regex matches, you can recall the captured matches many statements later, that is, you need not do the recall immediately following the statement containing the regex.

For example, the captured values in $1 and $2 can be accessed after the substitute command has finished, that is, in a subsequent line of Perl, such as the following.

```
$fullName = "Treponema pallidum";
$shortName = $fullName;

# keep the first letter of the Genus name
$shortName = ~ s/^(.)\S+ (.*)/$1.$2/;
        :
        :
print "First letter captured is: $1 \n";
```

First letter captured is: T

Capturing essential parts of your regex matches and recalling those matches later in your program will serve you well in many situations.

5 Using Perl to Do Calculations

In which some of the mysteries of calculating with Perl will be elucidated.

If you are almost correct you are a liability.

—Fred Kollett

Arithmetic in Perl is fairly intuitive in its use of symbols. There are all sorts of reasons for doing a little arithmetic on the fly in the midst of your Perl program. For example, you may wish to calculate:

- percentages of As, Cs, Gs, and Ts
- exact locations (such as starting base) of a sequence in respect to other sequences in a file
- the probability that a particular string will appear, given a certain A:C:G:T ratio
- conversions from one unit (nanometers) to another (Angstroms)
- other applications that will appear in subsequent chapters on an as-needed basis.

Note: Running statistical tests is an obvious application in bioinformatics, although we will not be making specific recommendations as to which statistical tests you need for your own analyses. Use a statistics book as a supplement. Perl provides you with the arithmetic you need to set up your equations.

5.1 Arithmetic Operators

Writing programs to perform mathematical operations is similar to using a hand-held calculator. Like a calculator, Perl includes the standard set of arithmetic operators—addition, subtraction, multiplication, and division—as well as additional functionality like exponentiation, modulo division, and a rich suite of mathematical functions (e.g., square root, absolute value, cosine, etc.). The syntax for the arithmetic symbols is what you might expect, although note the subtle changes such as the use of the asterisk symbol (*) for multiplication, the double asterisk for exponentiation (**), and the percent sign (%) for modulo division.

Arithmetic operation	Symbol
Exponentiation	**
Unary minus (e.g., −3)	−
Multiplication	*
Division	/
Modulo division	%
Addition	+
Subtraction	−

As in a natural language, knowing the tokens (symbols) and syntax does not ensure that a sentence will be semantically correct (meaningful) or unambiguous. Unlike humans, computers in general and programming languages in particular do not tolerate ambiguity. A mathematical expression that is ambiguous to a human will be interpreted by Perl in a strict fashion according to a set of rules. Before plunging into some of the details, the take-home story is that programmers must work hard to avoid ambiguity in their solutions, and simple techniques like the liberal use of parentheses go a long way toward eliminating any real or potentially ambiguous expressions.

Consider the following seemingly ambiguous (at least to us humans) Perl statement:

```
$x = 2 + 5 * 4; # seems ambiguous, no?
```

Is the result, stored in the variable $x, the answer 28 (2 + 5 is 7, then multiplied by 4) or is the answer 22 (5 times 4 is 20, then added with 2)? The answer of course depends on whether the computer does the addition first or the multiplication first. (In Perl, the answer is actually 22 because the rule is that without parentheses, multiplication occurs before addition). Without a rule as to which operator to perform first, the statement is indeed ambiguous. To eliminate the ambiguity, Perl and all other programming languages define a precedence or priority for the operators. For each Perl statement involving arithmetic operators, the precedence hierarchy of the operators is consulted and those with the highest priority are performed first. If more than one operator of the *same* precedence appears in the statement, then the associativity is consulted, where associativity dictates if the operators with equal precedence are handled in order from left to right or from right to left. The precedence and associativity of the arithmetic operators is shown in the following table. Those operators appearing in the first row of the table have the highest precedence and are performed first, followed by the operators of the second row, and so on. Although parentheses ( ) are not operators, their use for disambiguation is so critical that we include them here.

Associativity	Symbols
inside-out, i.e., inside pairs first, then continue working outward	( )
right to left	**
right to left	− (unary, negative)
left to right	* / %
left to right	+ −

5.1.1 Examples of Operators in Action

Some examples will help reinforce the rules that Perl follows when handling a statement involving these operators.

```
$x = 2 + 5 * 4;
# (22) multiplication before addition: 2 + (5 * 4)

$y = (2 + 5) * 4;
# (28) parentheses force addition first

$z = 3 * 5 * 2;
# (30) left-to-right: (3 * 5) * 2

$x = 2 ** 3 ** 2;
# (512) right-to-left: 2 ** (3 ** 2)

$y = 1 + 2 * 3 - 4;
# (3) means: (1 + (2 * 3)) - 4

$z = 5 - 4 + 3 * 2 / 1 ** 0;
# (7) means: (5 - 4) + ((3 * 2) / (1 ** 0))
```

Becoming comfortable with the operator precedence, associativity, and overall order of operations is a crucial element when writing good Perl, but do not be discouraged if this takes some practice. In this spirit, we offer a step-by-step hand trace of a relatively long arithmetic statement:

```
$z = 5 - 4 + 3 * 2 / 1 ** 0; # not a nice expression
```

Notice that this statement has five different arithmetic operators. Two sets of operators have equal precedence, so you must be careful to do each set in the correct left-to-right order. Referring to the table of arithmetic operator symbols, we first handle all expressions within parentheses. Because there are no parentheses, we move down the table to the operator with the next highest precedence, exponentiation (**). Exponentiation has right-to-left associativity, so we must start with the rightmost exponentiation. In this case, there is only one exponentiation, so that is performed (1**0). Recall that any number raised to the power of zero has the result one (1). In the hand trace below, notice that the expression (1**0) is now replaced with the resulting value of that expression. We recommend that you practice deciphering arithmetic expressions in this manner. Having completed all exponentiations, we move down the precedence hierarchy to the next operator, the unary minus. There are no instances of this operator, so we again move down to the next set of operators: (*, /, %, that is, multiplication, division, and modulo division). All three of these operators have the same precedence and a left-to-right associativity, thus (3 * 2) is computed *rather than* (2 / 1). Clearly, 3*2 computes to a six (6) and that result is used as the left operand to the division operator (6 / 1), which results in six (6). Having computed the first few operators and substituting their resulting value, we are left with the expression:

```
$z = 5 - 4 + 6 ;
```

Because we have completed all of the multiplication and division operators, we move down the hierarchy again. Next, addition and subtraction are of the same precedence with a left-to-right associativity, thus the subtraction $(5 - 4)$ is computed next rather than $(4 + 6)$. Notice that in this case the left-to-right associativity makes a difference since $((5 - 4)+6) ? (5 - (4 + 6))$. Once all arithmetic is completed on the right-hand side of the statement, the final result of seven (7) is finally assigned into the variable $z.

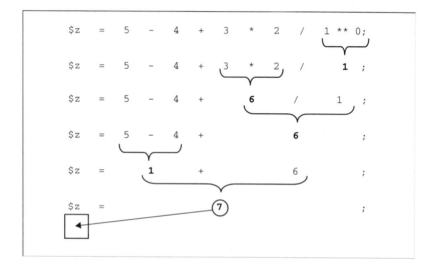

5.1.2 Unary Minus

The unary minus (negation) operator is somewhat of an anomaly. Intuitively, we use it all the time, for example, when we write "negative five," -5. Specifically, this negation operator is unary because it operates on only *one* operand, that is, it negates the sign of the value to the right. This is as opposed to the typical binary operators like subtraction that operate on two operands. Like exponentiation, unary operators are handled from right to left. Thus, you must handle each unary operator that you encounter in each statement starting on the right and working toward the left.

```
$x = -5;
# x now holds a negative five (-5)

$y = -$x;
# y holds negated value in $x, thus -(-5) or (positive) 5

$z = -$x + 1;
# x holds -5
# z now holds -(-5) + 1 or six (6)
```

Because the precedence hierarchy is not something that is easy to recall, programmers often get lax and enter their arithmetic statements as they would do algebraically on paper. But of course, Perl *never* forgets the correct order of operations, so a small error on your part can lead to wildly incorrect results.

```
# raise negative two(-2) to the fourth power (the wrong way)
$z = -2 ** 4; # WRONG! ** before unary(-), so does -(2**4)
# z holds -16 (not +16)!

# raise negative two(-2) to the fourth power (the right way)
$z = (-2) ** 4; # RIGHT!
```

In this example, exponentiation comes before unary minus, so if you do not use parentheses to force the unary minus to do its work first, you end up with a result of -16 rather than positive 16! Note that the lack of parentheses is not *wrong* as far as Perl is concerned. You should not expect Perl to halt your program and give you an error message. The statement is syntactically correct (and that is all the interpreter checks), but of course it is logically wrong. This is not cause for great alarm if you are writing small Perl programs on your own desktop computer; however, if this line of code is part of a system that is dripping medicine into your arm, we can agree that -16 units of medicine is not a good thing.

....

Good Practices

(Always use parentheses!) From the outset, when you are entering mathematical equations that involve more than one operator, we recommend that you get in the habit of using parentheses to surround each set of (operand operator operand). Although it is very important that you learn the order of operations so you can correctly interpret Perl statements written by others or even your own, this is one area where humans succumb to the unforgiving but consistent nature of machines. A good practice when dealing with complicated mathematical equations is to write down the formula on paper and then insert parentheses around the operations you expect to occur first, working outward surrounding combined operations within parentheses, and so on.

```
$z = $a - 4 + $b * 2 / $c ** $d; # ambiguous to a human!
$z = $a - 4 + ( ($b * 2) / ($c ** $d) ); # much better
```

Two important features of writing good programs are readability and maintainability. Readability means that a program is easy to interpret on a statement-by-statement basis. The programmer who considers readability is ever conscious of a future programmer who will be reviewing the program at a later time. Often the future programmer is yourself! Maintainability goes hand-in-hand with readability: a readable program is a program that is easily altered, updated, and maintained.

....

5.1.3 Modulo Division

The modulo division operator (%) may be new to you because this is not an operator you find on most calculators; however, modulo division can be extremely timely in a number of instances and is worth exploring briefly. Modulo division returns the "remainder" of a division.

$$\frac{\quad 1\ R\ 2\quad}{3\,|\,5}$$

Thus 5 % 3 (read "five mod three") is 2. Modulo is valuable when you want to do something for, say, 1000 times, and then restart your count back at zero each time you reach 1000. For example, assume you are sampling (and counting as you go) every 1,000 nucleotides in a downstream region. In this case, if you use ($count % 1000) the result of "any count mod 1000" will return a result in the range [0 ... 999].

if $count is	$count % 1000
0	0
1	1
:	:
999	999
1000	0
1001	1
:	:

In the next chapter we will discuss how a loop will enable you to maintain a running tally like $count. For now, practice with a few modulo operations and keep this section in mind as you move on.

....

Going Back for More

1. Create a small Perl program to test each one of the arithmetic operations. Before you run your program, solve the operations by hand first and then check to see if your Perl program agrees with your math. For example:

```
$a = 3 / 5;
$b = 5 / 3;
$c = 3.0 / 5;
$d = 5 / 2.51;

print "a: $a b: $b c: $c d: $d \n\n";

$a = 3 % 5;
$b = 5 % 3;
$c = 2 ** 6;
$d = -2 ** 6;

print "a: $a b: $b c: $c d: $d \n\n";
```

2. Alter your Perl program to generate some long "messy" expressions, solve those expressions by hand first, and then test yourself to see if you have handled the order of operations correctly. Show your "messy" expression to a fellow programming friend. Take special care to notice how they work as they attempt to interpret your expression.
3. Once you have finished with (2), add parentheses to your expressions to make them more readable. Do you get the same answers as in (2)?

....

Box 5.1 Napkin Math

No matter how clear you believe your math will be, do yourself a favor.

1. Go get (another) cup of coffee (or your favorite caffeine-rich beverage).
2. Take a napkin.
3. Write out your equations by hand on the napkin and solve them.
4. That way you'll be all set to check your numbers, the first time you click Run.

5.2 Mathematical Functions

Like Perl's suite of built-in functions that operate on strings, there is a library of mathematical functions in Perl that mirror the buttons you'd find on a good calculator.

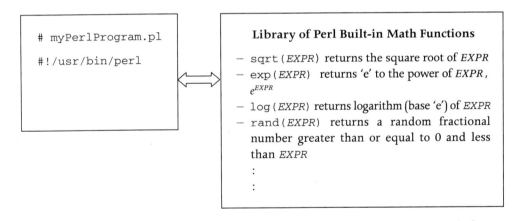

FYI: In chapter 14 we will cover how to add modules that contain additional functions that perform an even wider range of mathematical functions, for example, we will show you how to install a Statistics module (Statistics::Lite) that will extend Perl's built-in (default list) of mathematical functions to include a common set of statistical functions.

5.2.1 An Example Using `log()` and `exp()`: Finding the Likelihood of a Motif

One of the tools that our genomics research group has built is a motif lexicon or "DNA dictionary" (Dyer et al., 2004). The user selects a genome and enters a motif of interest, for instance, ACGT, and the motif lexicon returns a rich set of annotation about that motif including all locations (bp), links to upstream and downstream genes for each location, and introductory statistical information about the likelihood of finding this motif. The following example highlights the use of the mathematical functions `log()` and `exp()` in the context of computing how likely it is to find a motif of interest when working with a specific A-C-G-T distribution.

As shown in chapter 3, the transliterate (`tr`) function can be used to count the number of nucleotides in a sequence.

```
# determine the length of the DNA sequence
$sequenceLength = length($sequence);

# find number of Adenines (A), etc...
$numA = ($sequence =~ tr/A/A/);
$numC = ($sequence =~ tr/C/C/);
$numG = ($sequence =~ tr/G/G/);
$numT = ($sequence =~ tr/T/T/);
```

Using these newly computed frequency counts and the overall length of the sequence, one can easily compute the proportions:

```
# calculate the proportion of each nucleotide here
$probOfA = $numA / $sequenceLength;
$probOfC = $numC / $sequenceLength;
$probOfG = $numG / $sequenceLength;
$probOfT = $numT / $sequenceLength;
```

Given the nucleotide distributions and making the assumption that the nucleotides are independent from one another, one can use the product rule to easily compute the likelihood of encountering a particular motif. (*Note:* Assuming independence of nucleotides is done here to keep the mathematics simple. A more thorough treatment of this topic can be found in Durbin et al., 1998). Assuming Chargaff's Rule [p(A) = p(T) and p(G) = p(C)] is strictly applied in the following A-C-G-T distributions, the likelihood of the largely six-letter GC-rich motif GCCAGC as calculated by hand is:

$p(A) = 0.22$
$p(T) = 0.22$
$p(C) = 0.28$
$p(G) = 0.28$

considering a string of length 6 only

$$
\begin{aligned}
p(GCCAGC) &= p(G) \cdot p(C) \cdot p(C) \cdot p(A) \cdot p(G) \cdot p(C) \\
&= (.28)(.28)(.28)(.22)(.28)(.28) \\
&= 0.000378628 \\
&= 3.79 \times 10^{-4}
\end{aligned}
$$

```
# now done in Perl ...

# $probA is p(A), etc
#    :
#    :
my $motifProb;

# assuming independence of nucleotides
$motifProb = $probG * $probC * $probC * $probA
                    * $probG * $probC;

print "p(GCCAGC) = $motifProb \n\n";
```

```
p(A) = 0.22
p(T) = 0.22
p(C) = 0.28
p(G) = 0.28
------------------------------------------------
p(GCCAGC) = 0.000378628096
```

In this solution, a number of fairly small values are multiplied together to produce an even smaller value. In the spirit of the quote that leads this chapter, it is important to acknowledge the limitations of storing very small (or very large) values in a computer. Computers have a physical limit on how small and how large values can be to be safely represented internally. Numbers that get too large cause values to *overflow* (become larger than the largest possible value that can be stored on that machine) and values that get too small cause values to *underflow* (become smaller than the smallest possible value that can be stored on that machine). Overflow and underflow are difficult to detect, and therefore it is best to consider strategies for avoiding the situation when possible or needed. In the example here, multiplying many very small values together can potentially result in underflow. To avoid multiplying many small values together, we use logarithms to convert a lengthy series of multiplications into a series of sums. Avoiding multiplications of small values can help prevent the chance of underflow. Logarithms can help you convert a product of values into a sum of values, as shown here:

$$p(GCCAGC) = p(G){\cdot}p(C){\cdot}p(C){\cdot}p(A){\cdot}p(G){\cdot}p(C)$$

$\log(p(GCCAGC)) = \log(\ p(G)\ p(C)\ p(C)\ p(A)\ p(G)\ p(C)\ )$ *take the log of both sides avoid multiplying lots of small values together by noting that the log of the product is the same as the sum of logs of multiplicands*

$$sumOfLogs = \log(p(G)) + \log(p(C)) + \log(p(C)) + \log(p(A)) + \log(p(G)) + \log(p(C))$$

$p(GCCAGC) = e^{sumOfLogs}$ *take e^x of the sum of logs to remove the logs*

```
# alternative solution for likelihood of a motif using logs
# rather than multiply many small Real numbers together,
# use the fact that the log of the product is the sum of
# the individual logs of the multiplicands
# $probA is p(A), etc
# :
# :
my $motifProb;
my $sumOfLogs;

$sumOfLogs =  log($probG) + log($probC) + log($probC)
               + log($probA) + log($probG) + log($probC);

$motifProb = exp($sumOfLogs);

print "Using log and exp, p(GCCAGC) = $motifProb";
```

```
p(A) = 0.22
p(T) = 0.22
p(C) = 0.28
p(G) = 0.28
```

```
Using log and exp, p(GCCAGC) = 0.000378628096
```

Box 5.2 A Bookshelf of DNA

....

Consider a bookshelf holding all of the books needed to present the 3 billion bases of the human genome in a readable font. If each line of a page of a book contains 50 characters and if there are 30 lines per page, then 1,500 letters (A,C,G,T) can be recorded per page. If each book has 300 pages, then we have the space for 450,000 letters per book. That means we would need about 6,666 books. If each book were 2 centimeters thick, we would need a bookshelf of about 13,333 centimeters or 133 meters long.

Now, take that same readable font, with each 50-character line occupying 10 centimeters. How long a string would that be? Calculate it, or better still write a Perl script to calculate it for a genome of any size. (*Hint:* This is going to be a string of geographical proportions. See the introductory chapter for a similar calculation based on the molecular size, rather than the font size, of DNA.)

5.3 Formatted Output with `printf`

In general, the `print` function will suffice when you need to print some results or a message. However, there are times when you need more control of the way your output appears, such as when you are printing results of computations and your numbers appear with an unnecessary (and inaccurate!) significant number of digits. For example, one-third will appear as:

```
0.333333333333333
```

The `printf` statement, although somewhat cryptic, can facilitate the formatting of your output. This built-in function prints a **f**ormatted string by giving you more control over how you would like your real numbers, integers, and strings to appear. A quick example can reveal a primary reason you would use the `printf` function. The `printf` statement below prints one-third (`1/3`) in six spaces: two leading blank spaces, a zero, one of the total of six spaces taken up by the decimal point, and two (`2`) digits after the decimal point.

```
$a = 1 / 3;
print  "Unformatted: $a \n";
printf "Using printf: %6.2f \n", $a;
```

```
Unformatted:   0.333333333333333
Using printf:  0.33
```

printf "FORMAT", *VARIABLES*

The `printf` function requires two arguments: (1) a format string that dictates the width and precision (for real numbers) to use for each of the values to be printed and (2) the variables that hold the values to be printed. Variables that hold integers are printed with a `%d` format, variables that hold strings are printed with a `%s` format, and real numbers are printed with a `%f` format. Each format letter (`d`, `s`, `f`) may also be preceded by an optional width (integer) that dictates the number of places to use when printing the value. A positive width means to print the value within the width right-justified and a negative width means to print the value left-justified. In addition, real number formats allow an additional option to indicate the number of places to print after the decimal point. The following table and example summarize the available formats and show some of the more common uses.

When you need to format	In general	Example
a real number	**%**<*width*>.<*precision*>**f**	`%6.2f`
an integer	% <*width*>**d**	`%8d`
a string	% <*width*>**s**	`%25s`

```perl
#!/usr/bin/perl

use strict;
use warnings;

my $directoryName      = "Bacillus_anthracis_Ames_0581";
my $GC_ratio           = 0.35179;
my $numberOfPlasmids   = 2;

# %35s right-justify string in 35 places
printf "Directory Name: [ %35s ] \n", $directoryName;

# %-35s left-justify string in 35 places
printf "Directory Name: [ %-35s ] \n\n\n", $directoryName;

# %6.3f three-places after the decimal point
printf "GC_ratio: [ %6.3f ] \n", $GC_ratio;

# %6.1f one-place after the decimal point
printf "GC_ratio: [ %6.1f ] \n\n\n", $GC_ratio;

# %4d right-justify in four places
printf "Number of plasmids: [ %4d ] \n", $numberOfPlasmids;

# %-4d left-justify in four places
printf "Number of plasmids: [ %-4d ] \n", $numberOfPlasmids;
```

```
Directory Name: [        Bacillus_anthracis_Ames_0581 ]
Directory Name: [ Bacillus_anthracis_Ames_0581        ]

GC_ratio: [  0.352 ]
GC_ratio: [    0.4 ]

Number of plasmids: [    2 ]
Number of plasmids: [ 2    ]
```

You can use more than one format-variable pair in a printf statement. If you do, the first format matches the first variable, the second format matches the second variable, and so on, as shown here.

```
printf "GC_ratio: %6.3f - #plasmids: %d \n", $GC_ratio, $numberOfPlasmids;
```

A final note about the printf function is in order. You typically do not want to fuss too much about the way your output appears. As mentioned earlier, a good use of Perl is to use it to find your results, print those results separated by commas or tabs, and then format your output using another piece of software if you need it to look professional. For example, in chapter 8 we will show you how to print your results to a file. If you print tab-delimited output and name your output file with an .xls extension, you will be able to open your output with a spreadsheet program such as Excel and then use that program to make your output look nice.

Box 5.3 The Library of Babel

Now might be a great time to read or reread the wonderful short story by Jorge Luis Borges, "The Library of Babel." Various English translations exist that may be found online or at the library. The story consists of a detailed, often mathematical description of a fantastic library of all possible books (of length 410 pages) that can be created of 22 alphabet letters (the Spanish alphabet) plus a space, a comma, and a period for a total of 25 characters. "In the vast library, there are no two identical books." And the number of books "though extremely vast, is not infinite." Most of the books are either completely without meaning or nearly so. For example, "One ... was made up of the letters MCV, perversely repeated from the first line to the last."

Yet the library contains books on every conceivable subject as well as those that are inconceivable.

> The minutely detailed history of the future, the archangels' autobiographies, the faithful catalogues of the Library, thousands and thousands of false catalogues, the demonstration of the fallacy of those catalogues, the demonstration of the fallacy of the true catalogue, the Gnostic gospel of Basilides, the commentary on that gospel, the commentary on the commentary on that gospel, the true story of your death, the translation of every book in all languages, the interpolations of every book in all books.

As usual, we are thinking of DNA. Consisting of combinations of a mere four letters, the catalog (or library) of all possible DNA sequences (of gene length, for example) is vast magnitudes smaller than Borges's library but nonetheless daunting if analysis is the goal.

If we choose 3,000 bases as the length of a typical gene (producing a protein of about 1,000 amino acids) then the probability of any particular gene in the catalogue is 4 to the 3000th power (4^{3000}), a number far too great to have a name. True, we do not have to wade through all possible random and therefore meaningless sequences, but until we understand DNA coding completely, we will not always recognize the difference between noise and meaning. Natural selection sorts out things quite a lot for us, although selection also provides a loop of circular reasoning: The genes of successfully reproducing organisms must have sufficient meaning and a relative lack of random sequences to carry out the minimum necessary functions. That leaves the nontrivial task of analyzing what we mean by "sufficient," "relative lack," and "minimum necessary," which is much of what the work of DNA sequence analysis is about.

....

Going Back for More

1. Write a small Perl program that uses the square root function. Create a program that is easy to verify that you are using this function correctly.
2. Suppose you know the probability of getting an adenine (A) in a certain sequence of DNA. Using Chargaff's rules, write a Perl program to compute the other three probabilities. For example, suppose you *only* know that the probability of getting an A in the sequence is p(A) = 0.22.

```
my $probA;
$probA = 0.22;

# now compute the three other probabilities
```

3. Write a small Perl program to test the printf statement.
4. Assuming you know the A-C-G-T distributions in a DNA sequence, write a Perl program to compute the likelihood of the AT-rich (12-mer) motif TATAATATATAA.

....

Box 5.4 Joining the Programming Community or Creating Your Own Version of It

....

The field of computer programming is often associated with a sort of geek culture or community with its own acronym-based language and nocturnal customs. Although some of the extreme examples are mostly myths, elaborated by television and movie depictions of programmers, some are true and are reflected in the jargon and pastimes of folks who spend a great deal of time in front of a computer screen. If you are observing cautiously from a distance, perhaps trying to decide how much of geek culture, jargon, bravado, and work habits you would like to embrace, consider the following.

There is an opportunity here for you to own your own little piece of the programming community, create your own working conditions, and maybe even have a local influence on your programming colleagues. You may find acceptance of and even celebration of various forms of eccentricities by folks who are themselves somewhat extraordinary. Your own offbeat contributions may be welcomed or at least admired! One of the joys of embarking into a new field is the opportunity to reinvent yourself a bit and, in doing so, reinvent some of your surroundings. There is an enormous margin of error for what is acceptable in geek culture, as long as your program runs and gets results.

For example, your fellow programmers may be staying up all night drinking powerful, caffeinated beverages. You on the other hand may become known for your herbal tea prepared in a real china cup and your early bedtime, because most of your program got written before lunch while everyone else was still asleep.

Your fellow programmers might consider themselves to be "warriors," attacking problems by repeated assaults. They may spend their free time still in front of the screen in somewhat violent role-playing games. For exercise, they may keep squirt guns with which to ambush each other at 2 a.m.

However, you might use an old-fashioned fountain pen to write out all your algorithms ahead of time and at a picnic table outdoors, as far from the screen as you can get. (Of course you are not there at 2 a.m. to witness their reactions upon finding that the nozzles of the squirt guns are blocked with super glue.)

The other programmers compete with each other to write as few lines as possible and to use inside (slightly off-color) jokes to name variables. They might consider extensive comments to be for wimps. You write a dozen lines of code where six might be more typical, and you are known for your effusive, even poetic comments. As for variables, you don't get inside jokes and sometimes have to have them laboriously explained to you.

They wear T-shirts printed with undecipherable (to you) code; you have a bud vase attached to the side of your monitor.... You get the picture. Asking what a T shirt means is a conversation starter.

Suggestions for Creating and Expanding on Your Own Corner of Geek Culture

1. Deconstruct! Especially question acronyms and jargon on first usage. Cultivate an exceptional interest in having acronyms spelled out and then make a conscious decision as to whether the acronym is useful to you or whether you prefer the full name. Feel free to come up with your own terminology, for example, based on visual metaphors.
2. Ask questions that do not typically get asked. Essentially take nothing for granted. The brilliant Larry Wall invented Perl; the grammar and syntax have been elaborated over the years by many outstanding programmers. However, Perl is not a perfect language, and some of the decisions about Perl conventions may be either arbitrary or inflexible. It is okay to question some of the normal practices and to find out some of the historical basis for the way things are with Perl.
3. Come up with your own set of good practices with respect to
   - Naming variables, files, and programs
   - Writing comments
   - Using indentation to delineate parts of your code
4. More than just adopting good practices, you want to work on a personal style so that your variables and comments and formats are not merely logical. They should also have a particular hallmark that is yours. Style is something to be developed slowly and will not necessarily come out in your first few projects. Observe good programmers, ask them about decisions they make in constructing programs, and feel free to borrow some of their best ideas.
5. Your work style in general will most likely have deep roots from your years of handing in student assignments. Whatever your old habits are, they may be hard to break. Nevertheless, here is an opportunity to examine some of your typical practices as a novice in a new field. *Handling deadlines:* Will you be staying up all night or working little by little?

continued

5.4 Continued

> *Hacking:* Will you be trying bits of code repeatedly and running your
> program over and over with little tweaks, trying to figure out what is
> wrong? Or will you take a break from the screen and do some napkin
> math to try to figure out the bugz.
>
> *Reliance on other programs and programmers:* Quite a bit of Perl code can
> be lifted from other places and spliced into your program for
> your own purposes. Sometimes this makes much more sense than
> reinventing the wheel. The Perl community has a long, rich tradition
> of sharing snippets of code. On the other hand, you may end up
> spending just as much time on look-ups as writing a little function by
> yourself.
>
> *Collaboration:* Our reasonable guess about users of this book is that they
> are coming out of a tradition of collaboration, especially if they are
> biologists, and are looking for interdisciplinary interactions with the
> programming community. That is, you probably are not planning to
> be a reclusive, solitary hacker. Please see boxes throughout this book
> for suggestions about opening up communication in an interdiscipli-
> nary setting.
>
> 6. Your little niceties might include a comfortable chair, an attractive
>   screen saver, the right lighting and music, and going on vacations with-
>   out your laptop. Do not hesitate to provide those and more to yourself as
>   you expand on your own corner of geek culture.

6 : Making Decisions Over and Over Again with Perl's `if` and `while`

In which useful Perl "control structures" will be introduced allowing you to check for unique situations and relieving you of the necessity of issuing the same statements over and over. Sit back and relax while Perl does the work of methodically and exhaustively searching long strings of DNA sequences. Note that we revisit regular expressions.

The virtues of a programmer according to Larry Wall: "Laziness, impatience, hubris."

6.1 Simple Statements versus Controlled Statements

So far, you have been issuing simple commands in Perl, and as long as the spelling and syntax have been correct, you have been obeyed. For example, strings of DNA sequences have been sought, found, and manipulated in various ways. What are missing right now are certain subtleties that enable you to handle mutually exclusive events and relieve you of the chore of having to issue the same Perl statements over and over again.

You say, "Find this substring." But perhaps what you really wanted to say is: *"Find the motif ACGTGG and keep on finding it while you are still in the sequence and don't stop until you have found all of them."*

A "while loop" is what you need; it will issue your statements again and again "while" there is still more searching to do.

Or perhaps you would like Perl to check for a certain condition as it searches through a file of DNA sequence. You might want something like this: "Find a motif (substring) with certain characteristics. *If* it is of length greater than 10, then report it; if not (*else*) do not report it."

In programming parlance, testing for mutually exclusive situations and repeatedly executing commands are referred to as conditional and repetitional control, respectively. Most programming languages provide the programmer (you) with three types of control: (1) sequential (one instruction after another from top to bottom), (2) conditional (`if-else`), and (3) repetition (`while`). Up to this point, all your Perl programs have used only sequential control; that is, statements are executed sequentially, from the top of your Perl program to the bottom. This chapter introduces you to the other two types of control, conditional and repetition. All programs from this point will be creative blends of these three types of control structures.

6.2 **Conditional Control: Making Decisions with** `if-else`

Life and programming are full of opportunities for making decisions that are based on certain conditions. *If* I have a dollar or more, I can buy a cup of coffee or *else* if I have less than a dollar I cannot. That conditional statement can be written in Perl:

```
my $moneyInPocket;
$moneyInPocket = _____; # fill in an amount of money
if ($moneyInPocket >= 1)
{
    print "May I have a cup of coffee please? \n";
}
else
{
    print "I cannot afford coffee today. \n";
}
```

Conditional statements, often called if-else statements, have the following general form.

if (*a test which evaluates to true or false*)

> { *begin here if the test was true*
> > *do this*
> > *and this*
> } *end of true section*

else

> { *begin here if the test was false*
> > *do that*
> > *and that*
> } *end of false section*

The braces { and } form what is called a "code block," where all the statements within a block are treated as one unit. In the case of an if-else statement, the two blocks are mutually exclusive, that is, only one block of statements will be executed on a given run of the program depending on the result of the true/false test. If the test evaluates to true, then the if-block statements are run and the else-block is entirely skipped. Otherwise, if the test evaluates to false, the if-block is skipped and the else-block statements are run.

Notice in the example that the keyword *if* is followed by a logical comparison in parentheses (**$moneyInPocket** >= 1). When creating a logical comparison, the choices of logical operators are quite complete and mostly intuitive, although note the unique sets of operators to use when you are comparing numbers versus. when you are comparing strings. When comparing strings, Perl uses alphabetical ordering, that is, "A" is lt "C" and "Arginine" is gt than "Alanine."

Meaning	Use with numbers	Use with strings
Equal to (the same as)	==	eq
Not equal to	!=	ne
Less than	<	lt
Greater than	>	gt
Less than or equal to	<=	le
Greater than or equal to	>=	ge
Comparison	<=>	cmp

The comparison operators (<=> and cmp) will be covered later when we introduce sorting in chapter 9.

These collection of comparison operators are all you need to write if statements such as this one to test whether the length of a string is a certain value. In the example, length is a number, so a numerical operator is used, "==" (is exactly the same value as).

```
if ($lengthString == 10)
{
    print $string; # Print out only the 10-mers
}
```

However if you were comparing two strings, for example two sequences of DNA, the statement might look like the following. Note the correct use of eq because we are comparing two strings:

```
if ($string eq "actg" )
{
    print "I found actg.";
}
```

Notice that in these program segments, when the logical comparison is not true, then nothing is reported to you. Or if there is a solution to the problem that you have not considered (and therefore you did not include a logical test), the program has nothing to reply. Leaving your program like that means that it will be a little more difficult for you to sort out what might have happened when you test your program using various values or strings. For example, suppose that the sequence ggtg was stored in the variable $string. Clearly the string ggtg is not the same as (eq) to the string actg. Given these values, the eq test is "not true," thus your program replies nothing. It is silent. Hm, you wonder. Did the program actually work correctly and silence is actually a "reply" indicating the string was not matched, or have I made a mistake in my logic? Furthermore, for more complicated queries, you may not have thought of all possible answers. A good strategy is to allow your computer to reply "I did not find those particular matches," leaving you the option to consider whether other replies might have been possible. Adding an `else` statement will handle those contingencies and give your program a bit of finesse and even humility as shown below.

```perl
if ( $lengthString == 10 )
{
     print "Another 10-mer: $string \n";
}
else
{
     print "The sequence $string is NOT of length 10. \n";
}

if ( $string eq "actg" )
{
     print "I found actg. \n";
}
else
{
     print "The sequence $string does not match actg. \n";
}
```

The broad need for and use of if-else statements is evident when used in conjunction with some of the string functions that you studied in chapter 3. For example, the `index` function searches a string for a particular substring and if found returns the character position where the substring was found. When used in conjunction with the `index` function, you can print more specific messages based on the results, as shown here.

```perl
$sequence = "ACGTCGCCTATA"; # assume some unknown sequence
$sequenceLength = length($sequence);

# does the sequence end with "TATA" ?
$locationSTOP = index ( $sequence, "TATA");
if ( $locationSTOP == ($sequenceLength - 4) )
{
     print "$sequence does end with TATA \n";
}
else
{
     print "sequence does NOT end with TATA \n";
}
```

You may have noticed that we could have used the rindex function to search from the end of the $sequence, For example,

```
$locationSTOP = rindex ( $sequence, "TATA");
```

....

Good Practices

1. Programmers can be quite religious about the "correct" placement of begin { and end } braces in if-else, while, and other control structures. You'll note that we have chosen a practice of placing the braces on their own line and lined up.

```
if ( . . . . . . )
{
}
else
{
}
```

You will also see other styles, for example, the following.

```
if ( . . . . . . ) {
}
else {
}
```

No one style is best, but good programming demands that you and your group pick a preferred style and be consistent.

2. One of the most common sources of syntax errors for beginning programmers is mismatched parentheses and mismatched braces. Proper spacing and indentation of statements within blocks are two of the best ways to avoid these types of errors. We show the difference in the next two examples.

```
# bad style
if ($locationSTOP==($sequenceLength-4))          # needs spacing
{
print "$sequence does end with TATA \n ";        # needs indentation
}
else
{
print "$sequence does NOT end with TATA\n ";      # needs indentation
}
```

```
# good style:
#        braces { and } are lined up one above the other
#        use blank spaces before and after operators & parentheses
#        healthy dose of indentation within each block
if ( $locationSTOP == ($sequenceLength - 4) )
{
        print "$sequence does end with TATA \n";
}
else
{
        print "$sequence does NOT end with TATA \n";
}
```

Going Back for More

1. Write small Perl programs to practice with the logical operators for numbers (e.g., $<=$) as well as the operators used with strings (e.g., `le`).
2. Assume you have a string of three-letter amino acid symbols (e.g., `Lys` for lysine), for example:

   ```
   my $AA;
   $AA = "MetTyrAlaAsnAspIleValAsnSer...";
   ```

   Use the index string function in conjunction with an if-else statement to test:

   a. if the initial symbol is a Met
   b. if the ending symbol is a stop codon (indicated with ... here)

   Have your program print appropriate messages after testing each case.
3. Delete a brace } from one of your if-else blocks. Run the program and review the error message so you become familiar with the error message associated with this common mistake. After replacing the brace }, remove other braces and parentheses and review the associated error messages. Learning to recognize these types of error messages will serve you well in the future.

6.3 Testing for the Success of Finding Patterns: `if-else` and Regular Expressions

What about deciding *if* you have found a pattern when using a regular expression? This is the time to reintroduce regular expressions. Chapter 4 allowed you to try all sorts of creative variations for pattern matching using regular expression syntax. However, in that case you were focused primarily on learning the syntax for finding patterns. Now you are in a position to combine Perl's control structures with the power of pattern matching with regular expressions. Perl's match (**m**) operator, used in conjunction with the binding assignment operator (=~), allows you to search for patterns expressed as regular expressions. If a match is found, the if statement's logical test will evaluate to *true*, and if you "cluster" your regular expression within a set of parentheses you can "capture" the result of the match (e.g., a substring that matches the regular expression pattern) in Perl's global variable **$1**.

```perl
# search for a TATA-box
my $DNA;
my $TATAbox;

$DNA = "GCGACCACCTTGGTTCAGCAGTATAAAAAGGCGGCTTGGCG";

print "Assume we are searching the upstream region ";
print "of a certain gene.\n\n";

print "Region being searched is: $DNA \n\n";

print "=================================================\n";

if ($DNA =~ m/(TATA[AT]A[AT][AG])/)
{
        $TATAbox = $1;
        $TATAlen = length($TATAbox);
        $TATAlocation = index($DNA, $TATAbox);

        print "Found a TATA-box: $TATAbox \n";
        print "at location $TATAlocation \n\n";
}
else
{
        print "No TATA box was found \n";
}
```

Unix Script Output

Last Saved: 09/18/05 07:55:0፤
File Path: /Applications/BBEdit

Assume we are searching the upstream region of a certain gene.

Region being searched is: GCGACCACCTTGGTTCAGCAGTATAAAAAGGCGGCTTGGCG

===
Found a TATA-box: TATAAAAA
at location 21

6.4 Capturing and Clustering in Regular Expressions

As shown in the previous example, surrounding your regular expression with a set of parentheses is a way for Perl to "remember" the exact substring that successfully matched your pattern. Using parentheses in your regex is often called *clustering* and recalling the matched substring is called *capturing*. In chapter 4 when regular expressions were introduced, you used clustering *within* a regular expression to remember initial parts of a pattern when you needed to recall those parts of the pattern in the tail end of the regular expression. For example, recall our regex to find a direct repeat (DR):

> Are there any words in which a sequence of three letters is repeated elsewhere in the word?
> Search for words where a sequence of three letters is repeated.

```
(.)(.)(.).*\1\2\3
```

```
acclimatization
Albuquerque
amalgamate
  :
  :
```

In this example, the parentheses (.) were used to cluster (store) results for three consecutive letters and later in the same regex we used the back references \1, \2, and \3 to capture (recall) the first, second, and third characters previously remembered. But note that if we want to remember the entire matched pattern *outside* (after) the regex, we place a first set of parentheses around the entire regex and recall the match in a later line of Perl with $1. Thus a Perl if statement that attempts to find (match) direct repeats needs an additional set of parentheses around the entire regex and our back references within the regex must be renumbered. Starting from the left in a regex, each open parenthesis indicates the number that must be used to recall that particular item; the leftmost (first) parenthesis can be later recalled with (1), the

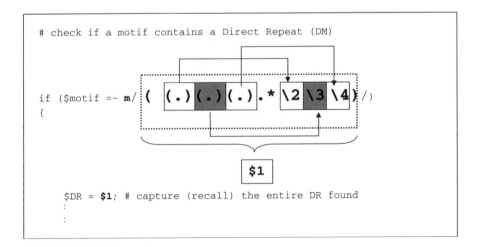

```
# check if a motif contains a Direct Repeat (DM)

if ($motif =~ m/ ( (.)(.)(.).*\2\3\4 ) /)
{

                          $1

    $DR = $1; # capture (recall) the entire DR found
    :
    :
```

second leftmost parenthesis with (2), and so on. Capturing previously clustered items within the same regex requires the backslash syntax (e.g., \2, \3), whereas capturing an item after (outside) the regex requires the typical Perl variable syntax expression, for example, $1, $2, $3.

Returning to the DR example, capturing (recalling) the second, third, and fourth items within the regex requires backslashes, and capturing the first (entire) match *after* the regex finishes requires **$1**.

EXPR =~ **m/***PATTERN***/** [option]

The **match** (m) operator searches a given string (EXPR) for a particular *PATTERN*, where the *PATTERN* is a regular expression. When used in the context of a logical expression such as within an if statement, the operator returns *true* if successful, *false* otherwise. A number of options are available when performing matches, for example /PATTERN/**i** when you wish to perform case-insensitive pattern matching. If you capture (enclose) the entire regex within a set of parentheses and a match is found, the result of the match is captured and stored in Perl's global variable **$1**.

....

Good Practices

When applying the match operator, the use of the letter **m** at the start of the match is optional; however, we strongly recommend that you include it. The short rule to follow is: "be explicit." The regular expression syntax is daunting enough on its own. The reader of your program (it may be you!) will appreciate the explicit reminder that this if-test is an attempt to match (m) a pattern (regex) to the string in question.

As always, use good variable names that will help the reader. When using Perl's numbered variables for capturing (recalling) the result of a regular expression, immediately store the numbered variable into a variable that defines the pattern you just matched.

```
if ($DNA =~ m/(TATA[AT]A[AT][AG])/)
{
    $TATAbox = $1; # immediately store with a good name
:
```

```
if ($DNA =~ m/((.)(.)(.).*\4\3\2)/)
{
    $MR = $1; # store the Mirror Repeat (MR) just matched
:
```

....

Notice that the use of regular expressions within an if-statement either succeed in matching or do not.

```
if ( $motif =~ m/((.)(.)\3\2)/ )
{
  $MR_4mer = $1; # store the entire 4-mer Mirror Repeat
  $MR_bp1 = $2;  # also have access to other captured parts
  $MR_bp2 = $3;
}
```

There is no need to use == or eq or any of the logical operators.

Going Back for More

1. Perhaps the regular expression patterns we've just used, for example:

   (TATA[AT]A[AT][AG])

   ((.)(.)\3\2)

   have caused you to say, "Whoa! What is **((.)(.)\3\2)**? Where did this strange syntax come from?" If so, now is a good time to look back at chapter 4. Regular expressions are powerful pattern matchers; **((.)(.)\3\2)** happens to find all possible mirror repeats of length four, such as CGGC, and cluster the entire match for later recall in $1. We are the first to acknowledge that there is a (serious) learning curve with regex syntax, and you should not be frustrated. That is why returning to chapter 4 to play around with English examples may be a good refresher. Then go ahead and insert your regular expressions into if statements using Perl's match (m) operator.

2. Write small Perl programs to practice with regular expressions in the context of an if statement and the match (m) operator. For example:

```
# find a 4-mer repeated, otherwise, alert user.
if ( $seq =~ m/(([ACTG]{4})\2)/ )
{
        $DR_4mer = $1;
        print "I found a 4mer repeated: $DR_4mer";
}
else
{
        print "no 4-mers repeated were found \n";
}
```

3. Determine if a DNA upstream sequence contains a TATA box. If it does not, report that and end the program. Otherwise, report its location and clip out the substring to the left of (upstream of) the TATA box. In the substring to the left of the TATA box, report *all* direct repeat (DR) motifs found with a total length of 4 to 12 bp and their location within the substring (e.g., GCGC, AAATTCAAATTC, etc.).

4. Add to your Perl from question 3 to search both upstream and downstream of the TATA box. Alter your program to search for other significant motifs of interest, For example, CAAT boxes, mirror repeats (MRs), and so on.

6.5 Logical Connectives (`and`, `or`, `not`)

In additional to the logical comparison operators (e.g., `==`, `!=`, eq, ne, etc.), Perl provides you with three logical connectors (`and`, `or`, `not`) that enable you to build more sophisticated if-else tests. For example, if you were checking for 10-mers *or* 12-mers, you could make the following test:

```
if ( ($lengthString == 10) or ($lengthString == 12) )
{
        print "Another 10-mer or 12-mer: $string \n";
}
else
{
        print "$string is NOT of length 10 or 12. \n";
}
```

When using logical or, only one of the tests need be true for the entire logical comparison in parentheses to result in true. In some cases, both tests may be true when using or, although in this example, they both can't be true. Why?

When you make a logical comparison with two tests and *both* the tests must be true for the entire comparison to be true, we use and. For example, if you wanted to find all sequences with lengths between 10 and 12, algebraically you would write:

$$10 \leq \text{length} \leq 12$$

Because we are asking two separate questions, "Is 10 less than or equal to the length?" and "Is the length less than or equal to 12?", we must use a logical and to connect these two separate tests.

```
if ( (10 <= $lengthString) and ($lengthString <= 12) )
{
        print "Length IS between 10 and 12, inclusive. \n";
}
else
{
        print "The length is NOT between 10 and 12. \n";
}
```

The following truth table summarizes the possible logical values (true or false) of each test and the logical result (true or false) for the entire test given each of the logical connectives. When connecting two tests with logical and, both tests must be true for the entire resulting expression to be true. When using logical or, at least one of the tests must be true for the resulting expression to be true.

Test 1	Logical connector	Test 2	Resulting expression
true	and	true	True
true	and	false	False
false	and	true	False
false	and	false	False
true	or	true	True
true	or	false	True
false	or	true	True
false	or	false	False
	not	true	False
	not	false	True

6.6 Testing for Multiple Cases: `if-elsif-elsif-else`

Sometimes you may want to string together a series of conditional tests that are all mutually exclusive. Although you could write a series of separate if-else statements, Perl has a control structure to facilitate the efficient testing of mutually exclusive situations. This is best shown by example.

```
# determine if the first codon (in the first Open Reading
# Frame, ORF) is one of the acidic amino acids (asp,
# aspartic or glu, glutamic) or a basic lysine (lys).
#       asp aspartic acid GAU or GAC
#       glu glutamic acid GAA or GAG
#       lys lysine basic AAA or AAG
#       otherwise, not an acidic AA or a basic lysine
my $RNA;
$RNA = "GAUAAGCGCCAC"; # assume RNA obtained somewhere
my $codon1;
# grab first codon
$codon1 = substr($RNA, 0, 3);

if ( ($codon1 eq "GAU") or ($codon1 eq "GAC") )
{
      print "Codon $codon1 asp (aspartic): ACIDIC \n";
}
elsif ( ($codon1 eq "GAA") or ($codon1 eq "GAG") )
{
      print "Codon $codon1 is glu (glutamic): ACIDIC \n";
}
elsif ( ($codon1 eq "AAA") or ($codon1 eq "AAG") )
{
      print "Codon $codon1 is lys (lysine): BASIC \n";
}
else # OTHERWISE
{
      print "Codon $codon1 is neither an acidic aa ";
      print "or a basic lysine. \n";
}
```

6.7 Avoiding Mathematical Irregularities

Perl's mathematical functions, like buttons on your calculator, assume that programmers like yourself will use them correctly. For example, we all know that we shouldn't try to take the square root of a negative number and we shouldn't try to divide by zero. Try these two operations on your calculator—what happens? Robust programs always check for situations that might cause problems. Checking for irregularities like dividing by zero in your program is called *trapping*, that is, you "trap" the bad situation with an if-else statement, avoiding the problem and reporting the trap to the person using your program.

A classic case in point is trapping division by zero in your programs. In many cases, your program handles a division operator just fine because the denominator is not (supposed to be) a zero. A robust program will trap a potential division by zero and offer the user (probably you) a nice message rather than allowing the division to occur.

```
# trap division by zero
if ( $totalNucleotides == 0)
{
        print "WARNING: trapped division by zero! \n";
        print "Denominator is: $totalNucleotides \n";
}
else # all is ok, denominator is not a zero
{
        $proportion_of_A = $numberOfAs / $totalNucleotides;
        print "Proportion of As: $proportion_of_A \n";
}
```

Going Back for More

1. Complete the `if-elsif-elsif . . . else` Perl program to print appropriate messages for all 20 of the amino acids.
2. Write a Perl program to obtain the length of a regulatory motif of DNA and print an appropriate message depending on the length of the regulatory sequence.

Range (bp)	Message to print
$1 \le \text{length} \le 4$	very small
$4 < \text{length} \le 8$	small
$8 < \text{length} \le 16$	medium
$16 < \text{length} \le 32$	large
otherwise	very large

3. A Perl program that divides by zero will terminate. Write a Perl program to force a division by zero and observe its behavior. Then, write an if-else statement to trap a potential division by zero and print an appropriate message, otherwise allow the division to proceed if the numerator is okay.
4. Write a Perl program to send a negative number to the mathematical square root (`sqrt`) function and observe its behavior. Write an if-else statement to trap a potentially bad negative value and print an appropriate message, otherwise allow the `sqrt()` function to proceed.

Box 6.1 Endless Loops and How to Stop Them

Before you get too far into while loops, take note of the way in which you will stop an endless loop if it should accidentally occur. Unlike other errors that might return erroneous or nonsensical answers, an error with a while loop might run itself in circles indefinitely, returning no answer, and possibly not allowing you to gracefully exit the program. As with any hung-up computer (including hung up at some Internet site), you need the correct magic words (combination of keys) for your operating system to stop the cycling. You might jot those down on a sticky note and stick them to your monitor!

Note: these are neither graceful nor subtle, but (alas) may be necessary if a loop has seized your program and is holding it hostage:

For Windows: Press Ctrl-Alt-Delete all together and then select the program you want to terminate.

For Mac: Press Option-Apple-Escape all together and then select the application you would like to quit.

However, you can avoid all sorts of problems with while loops and other control statements by:

1. counting all of your pairs of parentheses and brackets and making sure that they are indeed paired.
2. using a counter and checking its increment (or decrement) and halting test carefully.
3. stepping through the code before you run it—trying out all of your logical possibilities (if the statement is true, if it is false), by hand.
4. trying really tiny test files (or short test strings) to search, when you are piloting your program. It might be tempting to try a long sequence, knowing that Perl will blaze through with the while loop. However your program may return incorrect answers if some other bit of code is wrong, and it will be difficult for you to wade through that enormous string of DNA yourself trying to find the error. If your first test is on a short string, specially picked by you to challenge the program, you will be able to spot problems more quickly.

6.8 Repetitional Control: Repeating Statements Over and Over with `while`

Much of the same logic and syntax used in if-else statements can be used again in constructing `while` statements. The major difference between conditional control (if-else) and repetitional control (while) is that `while` statements allow you to build a loop that cycles through a section of your program as many times as requested. In short, `while` allows you to be relentless and thorough. `while` allows you to look and look through a long sequence of DNA (or entire genome) until either something is found or you are satisfied that it is not there. (*Note:* The previous programs using only if-else statements were content to stop when they had produced one [the first] answer and thus if-else statements were not obliged to go back and pull out any more.)

6.8.1 `while` and Counting

One challenging aspect of building a `while` loop is controlling it. You want the loop to continue when you have more work to do and stop when the work is done. In general, `while` loops are comprised of the five items indicated below.

[1] initialization
   (e.g., start your counter at one (1) and/or start your sum at zero (0))

`while ( [2] test` *some logical expression that evaluates to either True or False* `)`
`{`

> **[3]** *do this* **action** *when the logical expression evaluates to* **true** ;
>   *when true, we do these statements "inside" the while loop*
>
>   :
>
>   :
>
> **[4] update** *at bottom of the loop (before returning to the top of the loop)*
>   *we update a counter or perform some statement that*
>   *brings the logical expression "closer" to becoming* **false**

`} # end of the while loop` *(return to the top of the loop)*

[5] Continue *below the loop when the while loop test is **false***
   :
   :

One common practice is to use a `while` loop to repeat a set of actions a specific number of times. For demonstration purposes, assume you need to print a countdown from ten (10) down to one (1) and then print "Blastoff." As a preliminary step to determining the five items you need to build this loop, contemplate the set of instructions you would use if you did not have a while loop.

```
my $count;

$count = 10;              initialize counter              OUTPUT

print "$count \n";        print 10                        10
$count = $count - 1;      reduce counter by one to 9

print "count \n";         print 9                          9
$count = $count - 1;      reduce counter by one to 8

:
#(and so on)                                               :
:                                                          :
```

`print "count \n";`	*print 2*	**2**
`$count = $count - 1;`	*reduce counter by one to 1*	
`print "count \n";`	*print 1*	**1**
`$count = $count - 1;`	*reduce counter by one to 0*	
`print "Blastoff! \n";`		**Blastoff**

In practice, there is nothing to stop you from entering this set of statements. Clearly, it will work and it will produce the output that you want, until your colleague says, "Cool, can you make it print the countdown from 100 down to 1? How about 1,000 down to 1?" Two solutions might come to mind: (1) Copy and paste (no! don't do it!) or (2) a `while` loop.

Assuming you see the wisdom in the second solution, review the brute force solution above and see if you can spot the two instructions that are repeated verbatim over and over. The instructions that get repeated are exactly the instructions that belong inside your `while` loop. The instruction at the top is the initialization (because we only want to do this once before the loop begins) and the instruction at the bottom ("Blastoff!") we want to happen only once after the loop stops. Thus far, our `while` loop looks like:

```
$count = 10;
while ( $count >= 1 )
{
        print "$count \n";
        $count = $count - 1;
}
print "Blastoff! \n";
```

Consider the logical expression between the parentheses of the `while` loop. Remember, we want the loop to continue and do the actions inside the loop *when* the test evaluates to *true*, and we want the loop to stop and proceed with the statements *after* the loop when the test evaluates to *false*.

A common practice when analyzing a section of a program is to hand trace the statements of interest, where "hand trace" means you (the programmer) pretend you are a computer and you execute the instructions one at a time, keeping track of the variables and their associated values as you try each statement by hand.

At the start, the variable `$count` will be set to a ten (10), thus the test will evaluate to true because (10 `>=` 1) is a true expression. After one iteration of (one passthrough) the actions inside the `while` loop, a "10" will be printed and the variable $count will be decremented by one and will hold the value nine (9). Reaching the bottom of the while loop will automatically return us to the top of the loop (*note:* we will never return back to the statement $count = 10,only to the while test). Now $count is a 9, so the test is still true (9 >= 1). This continues until we print a one (1), decrement the $count to a zero (0), and retrn to the while test. The test is evaluated again, but now the result is false because $count is a 0 and the expression (0 >= 1) is a false statement. As described earlier, when the while test evaluates to false, the loop stops and control of the program continues below the while loop to the statement that outputs "Blastoff!"

6.8.2 `while` and Regex: Searching for All GC-Rich Motifs

While loops can be used to repeatedly search for matches (m) in strings. Previously, we used if-else statements to search for (one) match: Is it there, or is it not? But what if you want to continue searching for other matches, for example, downstream in the sequence after you find the first and each match? The while statement in conjunction with the match operator is what we want.

In general, repeatedly searching for matches in a string uses the following algorithm (recipe), presented here in pseudo-code (that is, not quite English, not quite Perl).

```
while (more DNA to search, look for the next match)
{
    # ok, we found a match
    (a) save the match
    (b) note the location where it was found

} # end of loop, return above to while statement
    # but begin next search after the match just found
```

This algorithm translated into Perl, in this case searching for patterns of GC-rich regions of length four to seven base pairs, might look like the following.

```
$DNA = "GCGAATATGGCGGGTCATCGCGGCGCCTGCCCATGTACG";
print "Region being searched is: \n$DNA \n";
print "0123456789012345678901234567890123456789012345678 \n\n";

# search for GC-rich runs of length [4-7] bp
print "-----------------------\n";
print "Searching for GC-rich regions of length [4-7]\n\n";

my $GCmatch; # hold the next GC-rich region found
my $GCloc;   # hold the location of the next match
```

```
while ( $DNA =~ m/([GC]{4,7})/g ) # globally search
{
     $GCmatch = $1;
     $GCloc = pos($DNA)-length($GCmatch);
     print "GC-rich region: $GCmatch at location $GCloc. \n";
}
```

```
Region being searched is:
GCGAATATGGCGGGTCATCGCGGCGCCTGCCCATGTACG
0123456789012345678901234567890123456789
```

```
----------------------------------------------------
Searching for GC-rich regions of length [4-7]

GC-rich region: GGCGGG at location 8.
GC-rich region: CGCGGCG at location 18.
GC-rich region: GCCC at location 28.
```

A few points about this while loop are in order.

```
while ( $DNA =~ m/([GC]{4,7})/g ) # globally search
{
     $GCmatch = $1;
     $GCloc = pos($DNA)-length($GCmatch);
     :
     :
} # end of the while loop

# continue here when no more matches are found
```

a. We are searching for GC-rich motifs of lengths between 4 and 7. The regex

   [GC] {4,7}

   means to match nucleotides of either a G or a C for a minimum of four but no more than seven times.

b. Notice the global (g) option after the regular expression. This is needed to force the matches to start each subsequent search *after* the previous substring that matched the regex pattern. If you forget the global (g) option, the regex will start the search at the beginning of the string each time, obviously not what we want if we want to find all occurrences of the match.

c. Like the if-else statement that uses match (**m**), if a match is found, the expression evaluates to true and the execution of your program enters the while loop.

d. At the bottom of the while loop, indicated by the end brace **}**, your program will automatically return to the top of the loop and execute the match in the while statement again. When no match is found, the execution of your program continues below the end of the while loop.

e. When used in conjunction with the /g regular expression modifier, Perl's pos() function will return the location (bp) just *after* the match. Because pos() returns the location after the match, notice that we have subtracted the length of the substring that was found to report the *starting* position (bp) of the match. For example, if $GCmatch was "GGGC" (length 4 bp) and pos() returns a 27 (the bp location just after the ending C in "GGGC"), then the GC-rich 4-mer begins at location: pos($GCmatch) − length($GCmatch) = (27 − 4) = 23.

6.8.3 `while` and Regex: Searching for a Certain Number of Motifs

Another, perhaps more relevant example is to search for a particular pattern but no more than a certain number of matches. In the previous example we discussed a while loop that searched for patterns (some regular expression), continuing until it found no more of that pattern. But what if you wanted to search and find no more than three patterns? The following example uses a match (m) test *and* a $count test in your while loop expression so you can find continue trying to find another pattern as long as there are more to find and you have not found more than three so far.

```
my $GCmatch;

my $DNA = "ACCGGTCGCCCTACTCCGGTACGAGGGGCCTGGCCCTCCCCCGTA";

$count = 0; # initialize the counter (none so far)

while ( ($DNA =~ m/([GC]{4,7})/g) and ($count < 3) )
{
        $GCmatch = $1;              # save the 4-mer

        $count = $count + 1;       # found another one, count it

        print "Found another GC-rich 4-mer: $GCmatch \n";

} # end of the while loop

print "\n";
print "We found a total of $count GC-rich 4-mers. \n";
```

```
Found another GC-rich 4-mer: CCGG
Found another GC-rich 4-mer: CGCCC
Found another GC-rich 4-mer: CCGG

We found a total of 3 GC-rich 4-mers.
```

Notice that more than three GC-rich 4-mers are in the $DNA string, but once the $count reaches three (3), the second part of the test ($count < 3) becomes false, thus the entire and test becomes false and the loop stops. Notice that because we started the counter at zero the while loop test must be ($count < 3) and not ($count <= 3). Why? Starting a variable that counts the number of occurrences of some item at zero is a wise choice because if we do not find even one match (that is, we never enter the while loop because the match test is false on the first try), the count is holding the correct number of successful matches at the end of the loop. In the case when we find no matches, the count will be zero.

6.8.4 while and Keeping Running Sums

Another common need for while loops is to continue to find some pattern and also keep a running sum of some bit of information, for example, a running sum of the lengths of all AT-rich motifs found between the lengths of 4 and 8. Keeping a running sum is similar to a counter in that each iteration through a while loop you want to increase the sum by some amount; however, in the case of running sums, the new value that you add on to the sum each time will be greater than one.

For example, consider the situation where you found the following motifs and computed their individual lengths. Clearly, the sum of their lengths is 31 bp.

$ATmatch	length($ATmatch) bp
ATTATA	6
AATT	4
TATATATA	8
TATAA	5
TTTTAAAA	8
	31

The following code segment shows a while loop that will find AT-rich motifs of lengths four to eight base pairs and report the total sum of their lengths at the end of the loop.

```perl
$DNA = "ATTATACAATTGCGTACTATATATACCGTATAACGTTTTAAAAAG";
my $sum;
my $ATmatch;
my $ATlength;

$sum = 0; # initialize a running sum of lengths to ZERO

while ( $DNA =~ m/([AT]{4,8})/g )      # find next motif
{
        $ATmatch = $1;                 # save the AT-rich motif

        $ATlength = length($ATmatch);  # determine the length
```

```
    $sum = $sum + $ATlength;            # add on the new length

    print "Found another AT-rich motif: $ATmatch ";
    print "of length $ATlength \n";

} # end while loop

print "\n";
print "The sum of all the motif lengths is $sum \n";
```

```
Found another AT-rich motif: ATTATA of length 6
Found another AT-rich motif: AATT of length 4
Found another AT-rich motif: TATATATA of length 8
Found another AT-rich motif: TATAA of length 5
Found another AT-rich motif: TTTTAAAA of length 8

The sum of all the motif lengths is 31.
```

Going Back for More

1. Write a Perl program with a while loop to print a countdown from 10 down to 1 followed by the word "Blastoff."
2. Modify your countdown program in (1) above to print a countdown from 100 down to 1 followed by the word "Blastoff." *Note:* If you have done the while loop correctly, you should only have to make one small edit in your program, right?
3. Write a while loop to find and print *all* the GC-rich motifs of lengths 3 to 10 and print the sum of the lengths of those motifs at the end of the loop.
4. Modify the program in (3) to find *only* up to the first two (2) GC-rich motifs of lengths 3 to 10 and print the sum of the lengths of those motifs.
5. Modify the program in (4) to find the first five GC-rich motifs.
6. Modify the program in (5) to print the average length (bp) of the first four GC-rich motifs.

Box 6.2 Eric Davidson

Eric Davidson is among those biologists who use a programming metaphor as a working hypothesis for gene regulation. Davidson's model organism, the sea urchin, has had several of its regulatory pathways worked out, complete with features that look something like while, for each, and if statements. Or at least it is possible to diagram the pathways in that manner.

One way to think of the metaphor is this: Consider a single gene to be a simple command or imperative statement: "Make this product," for example, "Make rhodopsin." Notice that genes being used in that way, as simple commands, will not produce the necessary flexibility, specificity, and nuances for coordinated cell functions, especially in multicellular organisms. What is missing is grammar or control statements, such as what we are introducing in this chapter. Using this grammar metaphor, a diagram of a well-regulated rhodopsin "command" might look more like a little program like this:

If the cell is in the retina of the eye
And
If the cell is depleted of rhodopsin
  Make rhodopsin

  And *for each* rhodopsin, make the modifications or splices needed to
  produce a functional molecule
And *if* the cell is still depleted of rhodopsin, repeat this loop.
Else
Do not make Rhodopsin

And don't forget to put in all the semicolons!

So where are these control elements in an actual genome? Many hypothesize and have shown that short motifs upstream and downstream of genes and even thousands of base pairs away from genes form a sort of grammar or program for regulation. Proteins bind in a coordinated fashion to each other and to particular DNA motifs to influence whether and how a particular transcript will be made. The analogy with a computer program falls apart somewhat in that gene regulation seems to involve quite a bit of sloppiness, fuzziness, redundancy, and lack of perfect efficiency, none of which are well tolerated by computers. There are plenty of semicolons left out and the "program" still runs well enough, typical of any evolved system.

However, an interesting aspect of Perl as a programming language is that it was designed to have some characteristics of an evolving system. These include the fuzzy, sloppy, redundant, inefficient properties of evolving natural languages. Therefore some of the parallels between Perl language structure and DNA regulation may be worth considering as working metaphors and hypotheses for genome exploration.

7 Subroutines

In which you will be encouraged to write short, stand-alone sections of a program, subroutines, that may be called in use whenever needed.

For I saw before me in dark contours the beginning of a grammar of biology.
—Chargaff, 1971

A metaphor for subroutines goes something like this. You organize your notes on the topic of cellular metabolism. Maybe you are a student studying for an exam or a professor constructing a lecture on the topic. Either way, a subject as intricate as cellular metabolism can benefit from a decomposition into modular units. For example, electron transport by itself is worth understanding as a process with its own inputs and outputs. The concept (and diagram) of electron transport can be placed module-like into several different metabolisms. It is used in photosynthesis driven by light, it is used in respiration driven by the breakdown products of food molecules, and it is used in several unique bacterial metabolisms. Electron transport is a sort of subroutine (one of many) for explaining and describing metabolism. Where exactly electron transport appears in your lecture or study notes is a personal decision. However, making it a stand-alone topic that can be referred to in several different contexts is not a bad idea.

Most complex computer programs are broken down into functional sections called subroutines. In Perl, subroutines and functions are synonymous, although we will tend to refer to built-in utilities as functions (e.g., length, substr) and the utilities that you write as subroutines. Subroutines stand apart from the linear order of a program's flow. The code for a particular subroutine is often represented just once in a neat package at the bottom of the program (or in another file) and then called into action as needed. Furthermore, once a subroutine is written, it can be easily cut and pasted into other programs; it is modular. Examples of subroutines might include complement, which, when called, will render the opposite strand of sequence. Likewise, a countNucleotide subroutine might stand by at the bottom of the program until needed to count the number of instances of a particular nucleotide within a sequence.

A wise use of subroutines makes programs

1. more readable because a section of code can be written once in a subroutine and used over and over in various part of the program by simply calling (by name) for its use; a call to the subroutine thereby replaces the details with a name;
2. more concise because you only need to make little adjustments and repairs to the one section of your code that performs the details;
3. reusable because a subroutine that works in one program today can be used in another program tomorrow.

As you have seen in earlier chapters, Perl supports a large number of built-in functions (or subroutines). For example, you can call the function square root by typing `sqrt` in your program. The hidden calculations to perform a square root are done behind the scenes, yet the result is returned to your program. The various string analyses that have been introduced previously, such as `length`, substring (`substr`), and `index`, are all built-in Perl functions. The focus and strength of any programming language is in part due to the nature of its suite of built-in functions. Perl's popularity for handling strings is due to its strong, comprehensive, and efficient list of string-focused functions.

If Perl does not have the built-in function you are looking for, it may still exist, having been written and archived by some member of the Perl community. There are many Web sites devoted to the distribution of subroutines, written in response to particular needs. Each subroutine extends the Perl language, packaging behaviors that are labeled with meaningful symbolic names. Also available are packages of interrelated subroutines, called modules, which we will introduce later in chapter 14.

Finally, if neither Perl nor the Perl community has the built-in function or subroutine you are wishing for, you can write it yourself. In the long run, taking the time to write a subroutine is more efficient and much safer than pasting in the same piece of code over and over for some repetitive analysis.

7.1 How to Find and Use Preexisting Subroutines

One of the common questions concerning subroutines is, "How do I know if a subroutine has already been written?" It is a good question and one that we recommend that you ask often. Clearly, if a subroutine already exists and you find out, you avoid rewriting it. Chances are good that if it is a built-in function (or a function that exists in an external module that you can load into your program; see chapter 14), the function works correctly and has survived some review process that deemed it worth keeping. In fact, like the buttons on a scientific calculator, the built-in functions are built in because programmers like you appreciate the abstraction that furnishes the functionality without having to know the details.

In addition to this book and other more comprehensive volumes on Perl, you can search the web using the phrase "Perl functions by category" to find groups of functions that are related by their functionality, for example, "operate on scalars," "operate on numbers," and "regular expressions and pattern matching." Such a search will reveal a rich list of functions for you to consider, including but not limited to

Functions for scalars or strings
```
chomp, index, lc, length, reverse, substr, tr///, uc . . .
```

Regular expressions and pattern matching
```
m//, pos, s///, split, . . .
```

Numeric functions
```
abs, cos, exp, int, log, rand, sin, sqrt, srand
```

Box 7.1 Erwin Chargaff, the First Bioinformaticist (He Counted As, Cs, Gs, and Ts)

....

There are people who suffer from giddiness when watching the Milky Way for
a long time on a clear night. Such people should not look into the cell nucleus.
—Chargaff (1963, p. 90)

Erwin Chargaff (1905–2002) has been assigned a rather limited role in the various retellings of the history of DNA. However, giving him some amount of credit is nearly unavoidable, given his eponymous numbers, Chargaff's numbers, A being equal in frequency to T and C to G. The astounding significance of those equalities was noticed by Watson and Crick, only after numerous trial-and-error attempts at model building without Chargaff's numbers.

Erwin Chargaff is one of the often overlooked great thinkers in the history of DNA. He has a subtle legacy of influence in the unfolding of molecular biology. We consider him the first DNA bioinformaticist, that is, the first counter of As, Cs, Gs, and Ts. Just as important, he was one of the first philosophers of DNA, devoting serious contemplation and eloquent writing on the question of what the sequences could mean. He lived long enough to see many of his predictions come true. As early as 1947 he stated it was "clear that whatever 'code' was carried by the nucleic aids must be imprinted on the sequential arrangement" (1963, p. 133). This was well before the nature of that sequence had been determined. It was that idea (of information in the sequence) that prompted Chargaff's analysis leading to the publication of the Chargaffian ratios A:T and C:G and perhaps a near miss at deducing the rest of the significance.

In his *Essays on Nucleic Acids*, published in 1963 (10 years after the publications of Watson, Crick, Franklin, Wilkins et al.) with computation not yet applicable to the problem, Chargaff noted that although an alphabet of four or five letters may look meager, "the result may be oppressively informative." He added that even with "stepwise and orderly dismemberment," the labor would be so slow that the cryptographer would find that the sequence had evolved, and the analysis must begin again.

In that same book of essays, Chargaff calls on the "modern disciplines" of "cybernetics and information theory." It must have been gratifying to him to see in the last 20 years of his life the great strides in acquiring and analyzing sequences and making it all so accessible at NCBI.

7.2 Calling to and Returning from Subroutines: A Walkthrough

Subroutines are small, modular sets of statements that perform one specific task, typically a task that you need to repeat a number of times. A subroutine can hide the sometimes messy details of a particular task. Once written, the programmer need only call for the subroutine to do its work whenever needed, thereby avoiding a duplicate of details each time the task is called for.

```perl
#!/usr/bin/perl
use strict;
use warnings;

my $DNA = "CCGATGCTACGATTTCATTCAGGTC";

my $complement_DNA;

print "5' $DNA 3' \n\n"; # note the use of 5' and 3'

# call the subroutine with the $DNA string
$complement_DNA = complement( $DNA );

print "3' $complement_DNA 5' \n\n"; # note 3' and 5'

#-----------\
# complement \
#-----------------------------
# This subroutine accepts a sequence of DNA and returns
# the complement sequence of DNA.

sub complement
{
  my $dna = shift(@_);        # the arguments (values sent) are
                              # held on a list in Perl variable @_
                              # so we shift the first (and only)
                              # argument off that list and
                              # store it in this subroutine's $dna

  my $antiSense = $dna; # store sequence into a new variable

  # use a regex to transliterate into complement nucleotides
  $antiSense = ~ tr/ACGTacgt/TGCAtgca/;

  return $antiSense;
}
```

The following example shows a Perl program that includes a subroutine called complement that produces the complementary sequence of a given sequence of DNA. The output of the original sequence (5' to 3') printed before the call to the complement subroutine and the associated complementary sequence (3' to 5') printed after the call is shown below.

```
5' CCGATGCTACGATTTCATTCAGGTC 3'

3' GGCTACGATGCTAAAGTAAGTCCAG 5'
```

A careful walkthrough of this program will help explain how the Perl statements at the top communicate with the subroutine, in particular how (0) control proceeds as normal in your program, (1) the Perl statements at the top call the subroutine,

(2) the original DNA sequence is sent "down" to the subroutine and stored on Perl's internal list of values sent (that list is named @_), (3) the subroutine retrieves (shifts) the value off the list and stores the DNA sequence in a new variable, (4) the subroutine does its work of finding the complement sequence, (5) the subroutine returns an answer back to the calling statement, and (6) the statements at the top receive the result from the subroutine.

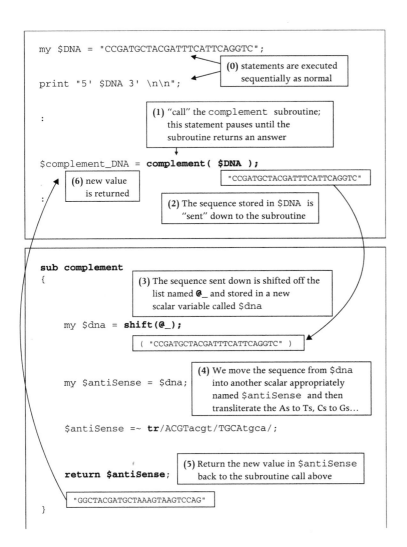

Notice that the subroutine complement is only a few lines of Perl, mainly the use of the tr operator to perform the necessary transliteration of characters. A common reaction is to ask, "Why write a subroutine? Wouldn't it be easier to just do the transliteration (tr) *in* the program?" Our answer is "no," for at least two reasons:

1. By writing the subroutine, the Perl program that calls the subroutine is very readable. Someone reading this program will immediately understand that finding the complement of a DNA sequence is involved. If they are curious and

want to know *how* a complement can be obtained, they can of course read the Perl statements inside the subroutine. At the time you are writing a program, readability might not be high on your list of priorities; however, from experience, we can assure you that writing programs that are readable saves time in the long run.

2. Once the complement subroutine has been written, whenever you need to take the complement of a sequence, you only need to call your subroutine. This holds true in this program and in any other program where you include the complement subroutine. So rather than rewrite the somewhat complicated transliterate (tr) statement over and over, you can insert the very readable subroutine call using different sequences as shown below.

```perl
my $mouse = "CCGATGCTACGATTTCATTCAGGTC";
my $human = "ACGATTTCATCCGGCGCGCCAGGTC";
my $yeast = "TGGTCCCGATGCTAGGGGGTCATTC";

my $complement_mouse;
my $complement_human;
my $complement_yeast;

# use the complement subroutine with various sequences
$complement_mouse = complement( $mouse );

$complement_human = complement( $human );
$complement_yeast = complement( $yeast );
```

Likewise, you could repeatedly call the complement subroutine from within a while loop. The following example repeatedly finds GC-rich 4–7-mers and then calls the subroutine to determine their complement sequences.

```perl
# find all GC-rich 4-7mers and determine their complements

my $GCmatch;

while ( $someDNA = ~ m/([GC]{4,7})/g ) # globally search
{
        $GCmatch = $1;

        print "5' $GCmatch 3' \n\n";

        $complement_DNA = complement( $GCmatch );

        print "3' $complement_DNA 5' \n\n";
}
```

Good Practices

It is considered wise programming practice to write documentation (# comments) for each of your subroutines. Your documentation should give a short SUMMARY of what the subroutine does (*not* how it does it), a description of each of the items that are sent IN to the subroutine, as well as a description of the value that is to RETURN. In addition to the English descriptions, we provide a sample style that makes your subroutines stand out as independent sections of code (e.g., like a file in a filing cabinet) to facilitate the copy-and-paste use of the subroutine in later programs.

```
#-----------\
# complement \
#-----------------------------------------------------
# SUMMARY: This subroutine accepts a sequence of DNA and
#          returns the complement sequence of DNA.
#
# IN: a sequence of DNA (string scalar)
#
# RETURNS: the complementary sequence (string scalar)

sub complement
{
    my $dna = shift(@_); # shift the string value sent
    my $antiSense = $dna;
    $antiSense =~ tr/ACGTacgt/TGCAtgca/;
    return $antiSense;
}
#-----------------------------------------------------
```

Box 7.2 The Etymology of the Metasyntactic Variable Foobar

We came across an amazingly exhaustive document from Donald Eastlake, Carl-Uno Manros, and Eric Raymond (The Internet Society) on the etymology of the word *foo*. (http://www.faqs.org/rfcs/rrfc3092.html). *Foo* and its extension *foobar* as well as the diminuitive *bar* are temporary default names for variables, introduced on the fly and before any thought can go into assigning a more useful name. Many computer scientists use foo and foobar without knowing much or anything about their history. First-time programmers may be puzzled by the sudden appearance of such metasyntactic (lacking proper definition) variables. Other such variables include qux, quux, grault, waldo, fred, and xyzzy.

Foo may have originated in the 1930s as a comic strip phrase, sounding vaguely Yiddish or even Chinese. For example, foo appeared often on signs and license plates in Bill Holman's "Smokey Stover" comic as well as in Walt Kelly's "Pogo." As a result, foo became a trendy slang word in the 1930s. Then it seems to have been picked up by the American military in WW II. There it was

continued

Box 7.2 Continued

re-employed in a subversive backronym: FUBAR for "Fouled Up [or worse] Beyond All Repair" along with related acronyms, such as SNAFU for "Situation Normal—All Fouled Up."

Apparently in postwar years, foo and its derivatives (including the acronym F.U.) were adopted by early programmers as useful technical jargon, couched in boyish insider humor. Early spellings were either foobar or fubar; one source of its spreading use in computing may have been manuals from the Digital Equipment Corporation in the 1960s, including the backronym use of fubar for "failed unibus address register."

Now that you know all of that, what might you (a relatively novice programmer) do? There are several choices:

1. Simply join the club and employ foo, foobar, and bar just like the experienced programmers do; that should include not leaving those names permanently in your code but using them only when you are sketching out ideas for code on the board and are at a temporary loss for a descriptive name.

2. Make up something more creative and more distinctive. In fact, you could come up with your own signature metasyntactic variable names, such as termite or blueCar and glueTar to rhyme with foobar.

3. Do your part to deconstruct technical jargon, especially when it is part of the network of insider (often male-oriented) humor that pervades computer science. Most programmers do not know the etymology of foo, so you can be the first to explain it in detail to them.

4. You could even begin a brave campaign to discourage the naming of variables and functions and passwords in code with hidden or thinly veiled sexist humor. Foo is actually just one mild example, especially with its comic strip origins; there is no logical reason to use four-letter words in code nor to hide their use in acronyms. Feel free to question that practice when you come across it.

By the way, biologists have not been shy about contributing their own insider jargon ranging from clever and delightful to unambiguously xenophobic and sexist. For a range of examples, try an Internet search on *Drosophila* (fruit fly) gene names.

7.3 More Examples of Subroutines

7.3.1 Counting Nucleotides

The following subroutine accepts two items, (1) a sequence of DNA and (2) a particular nucleotide, and returns the number of times that this nucleotide appears in the DNA sequence. This functionality is another good candidate for a subroutine because it is something you will often need but would like to avoid having to rewrite the details each time you need to count the number of times a particular nucleotide is present in a sequence.

```perl
my $numC;
my $numG;

$numC = countNucleotides( $DNA, "C" );

$numG = countNucleotides( $DNA, "G" );

print "$DNA \n";
print "Number of Cs: $numC \n";
print "Number of Gs: $numG \n\n";


#-----------------\
# countNucleotides \
#-------------------------------------------------------------
# Subroutine to COUNT and RETURN the number of nucleotides
# in the given sequence.
#
# two arguments: sequence to search through and the
#                nucleotide to count
#
# RETURNS: number of times the second argument appears in
#          the first arg sequence
#
# A sample of how you might "call" this subroutine from
# your program:
#
#      $numberOf_A = countNucleotides("ACGTACGT", "A");
#      print "$numberOf_A"; #--> would print two(2), two As

sub countNucleotides
{
  # list of arguments in @_ is: (sequence, nucleotide)
  my $someSequence = shift(@_); # shift off first argument
  my $whatBP       = shift(@_); # shift off second

  my $howMany; # total number of nucleotides found

  $howMany = 0;

  # count number of matches for the nucleotide in $whatBP
  while ( $someSequence = ~ m/$whatBP/g )
  {
    $howMany = $howMany + 1; # found another so count it
  }
  return $howMany;
}
#---- end of countNucleotides ---------------------------
```

The output of using `countNucleotides` is shown below.

```
CCGATGCTACGATTTCATTCAGGTC
Number of Cs: 7
Number of Gs: 5
```

7.3.2 Printing Sequences with Varying Widths per Line

Printing sequence to the screen is something we do often; however, the format for that sequence typically depends on the particular application. Often we want a standard 70 symbols per line, yet sometimes the number of characters per line varies, for example, 10 nucleotides per line, 20 amino acids per line, and so on. The following subroutine prints a sequence with a given number of characters per line. Notice in this example that no value is returned. This is perfectly acceptable. In this case, no `return` statement is needed at the end of the subroutine because the subroutine's job is only to print items.

```
my $odorant_receptor = "ggcacgagctggttccggaaagcctcatatctcgtatctt
                        aaagtat";

my $protein = "MTTSMQPSKYTGLVADLMPNIRAMKYSGLFMHNFTGGSAFMKKVYSSVH
               LVFLLMQF";

printSequence( $odorant_receptor, 10 );

print "\n";

printSequence( $protein, 20 );


#--------------\
# printSequence \
#-------------------------------------------------------------------
# Subroutine to print a sequence, $N characters per line
#
# IN: a sequence to print (could be DNA, RNA, or Amino Acid
#     sequence) and the number ($N) of characters to print per line
#
# RETURNS: nothing (ignored)
sub printSequence
{
  my $someSequence = shift(@_); # shift off first argument
  my $N            = shift(@_); # shift off second argument
```

```
    my $fromHere;              # next substr begins here
    my $nextLine;              # holds result of last substr
    my $lengthOfLastLine;      # length of last substr will be $N
                               # except last one

    $fromHere = 0;             # initial substr starts at the beginning
    $lengthOfLastLine = $N;    # allows control into loop the initial
                               # time
    # while substrings taken continue to be of size $N
    while ( $lengthOfLastLine == $N )
    {
        $nextLine = substr($someSequence, $fromHere, $N);
        print "$nextLine \n";

        $fromHere = $fromHere + $N;             # start of next
                                                # substr is here
        $lengthOfLastLine = length($nextLine);  # substr is how long?

    } # end while still more substrings to take

}
#--- end of printSequence-------------------------------------
```

The output after the two consecutive calls to the `printSequence` subroutine is shown here: 10 nucleotides per line and then 20 amino acids per line.

```
ggcacgagct
ggttccggaa
agcctcatat
ctcgtatctt
aaagtat

MTTSMQPSKYTGLVADLMPN
IRAMKVSGLFMHNFTGGSAF
MKKVYSSVHLVFLLMQF
```

Like with most solutions in Perl, there are many ways to do the same thing (see box 7.4). The repeated use of `substr` is but one way to write the `printSequence` subroutine. In the spirit of many ways, we present an alternative solution, `printSequence_2`, that uses a regular expression to match any $N number of characters each time. The regex, `/(.{1,$N})/`, matches any character (`.`) from one to $N times. Because regular expressions are "greedy" by default (see chapter 12), the regex will match the most number of characters whenever it can, thus except for the last match, it will match any N characters each time. The regex is surrounded by parentheses, so the resulting match is captured and can be recalled with the $1 variable inside the while loop. This subroutine produces the same result as the prior version.

```
# another way to write the printSequence subroutine

#----------------\
# printSequence_2 \
#-------------------------------------------------------------------
# Subroutine to print a sequence, $N characters per line
#
# IN: a sequence to print (could be DNA, RNA, or Amino Acid
# sequence) and the number ($N) of characters to print per line
#
# RETURNS: nothing (ignored)
sub printSequence_2
{
  my $someSequence = shift(@_); # shift off first argument
  my $N            = shift(@_); # shift off second argument

  # globally (g) match from 1 to $N of any character (.)
  # and print on new line

  while ( $someSequence =~ m/(.{1,$N})/g )
  {
      print "$1 \n";
  }

}
#--- end of printSequence_2-------------------------------------
```

Box 7.3 Perl Poetry

. . . .

It is well worth an Internet search to look up Perl poetry or Perl haikus. These stand-alone bits of code may or may not run. They are mostly playful metaphors in Perl syntax. Larry Wall is reported to have written the first one in 1990:

```
print STDOUT q
Just another Perl hacker,
unless $spring
```

These next two are from the Perl haiku contest 4.0; read them again after you have looked at chapter 9 on arrays (@) and chapter 10 on hashes (%).

```
By Ronald J. Kimball (who notes that summer.pl is a
sum-mer and a season)

sub summer { my $sum;
$sum += $_ for @_;
$sum } print summer (split);


By Steve Trigg (about a dying samurai)

unshift @pool, $sky, $clouds;
bless {sleep my $love and study};
each %willow and carp;
```

7.4 Terminology: Arguments

Subroutines are distinguished by their formal names. Arguments, sometimes called parameters, are the scalar values sent to a subroutine when that subroutine is called. For example,

```
length( $DNA );                    # 1 argument to length

printSequence($protein, 20);       # 2 arguments to
                                   # printSequence

index($putativeGene, "AC", 7);     # 3 arguments to index
```

The terminology becomes important as you read documentation for functions. For example, the documentation for the index function says that the first two arguments are required (the string to search and the substring to look for), whereas the third argument (where to begin the search) is optional.

Box 7.4 Tim Toady

Outside of idealized textbook illustrations, most cell structures and functions are remarkably like Rube Goldberg inventions or (our point throughout this book) like own natural language, fuzzy, convoluted, and good enough.

"There is more than one way to do it," says Larry Wall, and the phrase is so often used in Perl culture that the acronym TIMTOWTDI is pronounced *Tim Toady*. Tim Toady could be said of evolved systems, too. For example, Darwin noted that the finches of the Galápagos Islands had evolved to become pretty good woodpeckers and nut-eaters and blood-suckers in the absence of the

continued

Box 7.4 Continued

usual species that do those activities. The finch way was a little different, but good enough.

Some researchers have been tempted to make parallels between evolution and programming, such as Eric Davidson, who makes a reasonable analogy between programs and gene regulation (box 6.2). However, in general, computer programs are not especially good models for evolution. For example they are intolerant of spelling and punctuation errors. Try leaving out a single semicolon! Larry Wall, an avowed evolutionist, tried to include an evolution-like flexibility in the design of Perl. However, Perl is somewhat stuck in the world of perfectionism and does not approach true evolution for tolerance of variability.

All of this bodes for many years of hard work ahead on DNA sequence analysis. The fuzziness, redundancy, and noisiness of evolved systems make them especially difficult to decipher. A designed, computer program–like system would be easier to tackle!

7.5 Scope

In Perl and most other programming languages, the variables that you declare and the values stored in those variables are available for you, the programmer, to use within the block of Perl where each variable was declared. The specific block or blocks that each variable can be accessed is known as that variable's *scope*. Appreciating how scope fully works is, well, beyond the scope of this text; however, an introductory appreciation can help you avoid some messy errors as well as encourage you to follow some good programming practices that will serve you well in the future.

Recall that we have encouraged you to always use the following line at the top of all of your Perl programs.

```
use strict;
```

This directive to the Perl system requires you to declare every variable with the Perl keyword my, for example,

```
my $DNA;
my $len;
```

The keyword my declares a variable and defines its scope to be from the point of declaration until the end of a block unless that variable is redeclared. A block is delineated by Perl's begin and ending braces { }; if there is no enclosing set of braces, the scope is the entire (current) file. This all becomes important if you choose to use the same variable name in multiple places in your program. If you use the same name, the scope of the variable with multiple names determines which variable is actually being referenced. An example can help shed some light here.

```perl
#!/usr/bin/perl

use warnings;
use strict;   # you must use 'my' when declaring variables

my $DNA;  # has "global" (file) scope, unless redeclared
my $len;  # has "global" (file) scope, unless redeclared


$DNA = "GCTTCGAGCGCG";
$len = length($DNA);

print "Before call to subroutine the DNA is: $DNA \n";
print "Before call to subroutine the len is: $len \n";     1

printDNA($DNA);

print "After call to subroutine the DNA is: $DNA \n";
print "After call to subroutine the len is: $len \n";     4



sub printDNA
{
        my $seq = shift(@_);

        my $DNA;    # REDECLARES new variable; this $DNA now
                    # has a (new) LOCAL scope
                    # we didn't redeclare $len, so $len
                    # is the SAME $len as above

        $DNA = $seq;   # changes $DNA with LOCAL scope

        print "\n";                                          2
        print "   IN sub printDNA at start:      $DNA \n";
        print "   IN sub printDNA at start:      $len \n\n";

        $DNA = "TATA";        # changes $DNA with LOCAL scope
        $len = length($DNA);  # changes $len everywhere

        print "   IN sub printDNA after change:  $DNA \n";
        print "   IN sub printDNA after change:  $len \n\n";
                                                             3
}
```

scope of $DNA and $len unless redeclared

$DNA redeclared so new scope

In the following example, a programmer has declared a variable near the beginning called $DNA to hold some sequence data. The program includes a subroutine called printDNA, and the programmer has declared another variable called $DNA within the subroutine. As shown, the scope of the initial declarations of the two variables $DNA and $len begins at the point of declaration and ends with the bottom of the file. Thus, all lines of Perl from the point of declaration and after, including the subroutine printDNA, are within the scope of these variables and thus can be directly referenced. However, because a new variable also named $DNA has been declared inside the subroutine printDNA, the $DNA inside printDNA begins a new scope for that variable from the point of declaration until the end of the block, which is marked by the end of the subroutine printDNA in this case. The important point is that a change made to the value in the variable $DNA within the subroutine printDNA (within the scope of the redeclared $DNA) is known only within that scope! Put differently, if you alter the $DNA variable within the subroutine, you will not be changing the value of $DNA outside this scope, thus you will not be changing the value of the variable $DNA at the top of your program.

A careful hand trace of the entire example and the associated output will reveal some of the significant affects of scope. The numbered labels (1, 2, 3, and 4) shown in both the Perl code and the associated output indicate the order that the print statements are executed.

```
Before call to subroutine the DNA is: GCTTCGAGCGCG          1
Before call to subroutine the len is: 12

    IN sub printDNA at start:       GCTTCGAGCGCG          2
    IN sub printDNA at start:       12

    IN sub printDNA after change:   TATA                  3
    IN sub printDNA after change:   4

After call to subroutine the DNA is: GCTTCGAGCGCG          4
After call to subroutine the len is: 4
```

Notice in the next hand trace that within the subroutine printDNA the variable $DNA has been redeclared and thus creates a new scope from that declaration until the end of the subroutine block. Inside this scope, the variable $DNA is changed to TATA, yet once the printDNA subroutine ends, this (redeclared) $DNA variable leaves its scope. When control returns to the top, where the call to the subroutine was made, we have returned to the original scope, and thus the original value of $DNA remains the same as before the call to the subroutine.

On the other hand, notice that the variable $len has not been redeclared inside the subroutine. Yet because the subroutine is within the original scope of the variable $len declared at the top, the programmer can refer to $len within the subroutine. In this case however a change to the variable $len within the subroutine is a change to the one and only variable $len. (A variable like $len is often referred to as a *global variable* because when a variable is declared only once, often at the top of the program, it is then known throughout the file or globally.) Again, unless redeclared, a variable's scope extends from the point of declaration to the end of the respective block. In this case, $len was declared at the top of the file so the end of the block is the end of this particular file.

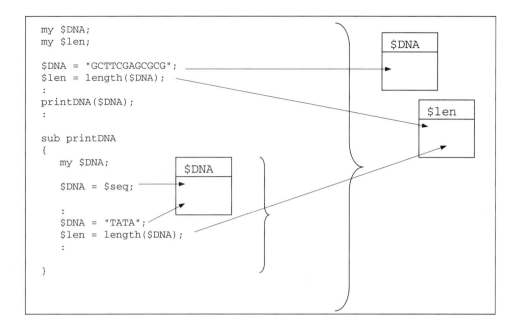

```
my $DNA;
my $len;

$DNA = "GCTTCGAGCGCG";
$len = length($DNA);
:
printDNA($DNA);
:

sub printDNA
{
    my $DNA;

    $DNA = $seq;


    :
    $DNA = "TATA";
    $len = length($DNA);
    :

}
```

Good Practices

We recommend the following practice when defining and testing a new subroutine.

a. Design your subroutine on paper. Carefully think about the task it will provide. Ask yourself: Is this subroutine performing one task? Or is it loaded down with extra work that perhaps should be broken down into more than one subroutine? Is the subroutine adequately named so that its behavior can be deduced from the name? (For example, a subroutine called loadAminoAcidTable is fairly obvious, while doNextItem may be clear to the programmer at one time, but not to anyone else or themselves at a later time.) Is the subroutine documented? That is, did you include some #comments to explain (i) a summary of the subroutine's purpose, (ii) the arguments that are needed, and (iii) the type of value returned to the calling program.

b. Are there any "tricky" side effects that may occur inside the subroutine statements? For example, it is often considered bad programming practice to alter a global variable inside a subroutine because the change is somewhat hidden from a programmer who is trying to decipher how a program is running. Although we won't go so far as to say you should never change a global variable inside a subroutine, we recommend that you carefully review how you have written each subroutine and take special care to document any such changes with a comment under the heading:

```
# SIDE EFFECT: changed global variable $len...
```

c. Enter your subroutine in a new Perl program. Add some additional lines of Perl to call your subroutine as a test.

d. Run this small Perl program to test your subroutine. Vary the arguments to your subroutine and test it on these values. Have you considered any special cases of arguments that might cause your subroutine to fail? (If so, you should document these cases in your subroutine documentation and/or write if-else traps at the top of your subroutine to generate warnings when appropriate.)

e. Once you believe you have thoroughly tested your new subroutine, you can integrate this subroutine into your larger program with some confidence that it has passed your tests and is ready to be used in larger contexts.

f. You may want to keep a list of "nice" subroutines that you write. Remember, one of the strengths of writing good subroutines is that you are writing software that can be reused, by other programmers and yourself!

Going Back for More

Write small Perl programs to practice with defining your own subroutines.

1. Implement one or both versions of the `printSequence` subroutine. Test it with various sequences.

2. Write a subroutine called `getSecondCodon` to accept one argument (a DNA sequence) and return the second codon in the initial reading frame of the coding region. The second codon is the second triplet in a DNA coding region, that is, assume the first occurrence of the three nucleotides (1, 2, and 3) ATG marks the first (start) codon and nucleotides (4, 5, and 6) denote the second codon. The DNA sequence might not start with ATG but you are assured that ATG does exist somewhere in the sequence, immediately followed by the second codon. The following Perl is a sample of a program that calls the function to test to see if the function works correctly.

```perl
my $DNA = "CCGATGCTACGATTTCATTCAGGTC";
my $codon2;
$codon2 = getSecondCodon( $DNA ); # call subroutine

print "The second codon is: $codon2 \n";
```

8 Accessing Files of Sequences from Databases

In which instructions are given for getting you to the vast databases of
NCBI and coming away with all of the files of sequence you want.

*[The London Library catalogue system] represents order—it is helpful, it leads
you to what you were trying to find, and also to what you needed, but did not
know you needed to find. It also has the delightfully mad quality of heterogeneous
things linked violently together by the arbitrary order of the alphabet.*

—Byatt, 2001

Sequence databases are the new museums of the genomic age. The grandest of them
all is publicly available at the National Center for Biotechnology Information (NCBI).
The databases at NCBI are among the best maintained, fastest, and most frequently
updated of any public databases in existence. Scientists worldwide are many times
fortunate that sequence information is so convenient and accessible and, most of
all, free. Nightmare scenarios could be imagined in which major governments had
declined to become involved with genomic sequencing and therefore most sequence
data were proprietary.

Some of the problems with sequence databases, even those at NCBI, are similar to
those encountered in the first museums dating from the seventeenth century.
Sometimes referred to as *Wunderkammers* (Chambers of Wonders), early museums
brought together diverse, curious, and rare objects often collected worldwide. How
to organize and arrange the artifacts was not at all obvious. The rocks, bones, feath-
ers, plants, holy relics, ancient coins, and much more (some items fragmentary) had
been collected well ahead of any knowledge as to how they might best be classified.

Such are the challenges at NCBI to which sequence data pours in various states of
annotation and completion. The overall design of a database requires great insight
and forethought as to what classifications should be used; this is particularly difficult when there is still so much about sequences that we do not understand. The
designs of databases to store genes have been fairly obvious, although not without
pitfalls as introns and exons are identified and their functions deciphered and
pseudogenes, horizontal transfer, and transposons must be dealt with. The classification of intergenic sequences is still in its inception because so little is known of
what the relationships and functions will be. Some aspects of intergenic regions may
be true mysteries (at least for us) in that they might not be deciphered in our lifetime.

An interesting case in point was the enormous influx of data entered at NCBI from
the *Beagle*-like expedition of J. Craig Venter and his team on the *Sorcerer II*. They
circumnavigated the globe, pulling up water samples about every 200 miles. From
those samples, the team sequenced all of the microbial DNA they could find. Most of
the sequences were new and not readily classified with the rest of the prokaryotes,

an indication of how little we know of the open ocean. Therefore sequences by the thousands received identifiers as vague as "Sargasso Sea," indicating where the DNA had been found, but nothing more. Soon after, researchers (including Venter's own colleagues) began to complain that most of their BLAST hits at NCBI were coming up with long, uninformative lists of "Unidentified Sargasso Sea Bacteria." A temporary solution was to segregate the *Sorcerer II* data from the usual BLAST search files at least until further information could be derived.

An intriguing philosophical issue encountered both at museums and at genomic databases is the lack of singularity of the items in storage. We know of course that a single stuffed and mounted three-toed sloth on display at the American Museum of Natural History is not "the" three-toed sloth, nor even the "average" three-toed sloth. It is just one specimen of an individual unlucky enough to be caught, stuffed, and mounted. Likewise if we find a genome for *Bradypus variegatus* (three-toed sloth) at NCBI, we should understand that it is not the genome but merely one variant. This point is made even more emphatically when we realize that there are individual humans and dogs (with names, personalities, and predispositions to diseases) who were the first to have their sequenced DNA achieved at NCBI. These individuals have come to represent a genome but certainly not the human and canine genomes at NCBI.

The lack of an Aristotelian ideal or universal essence of a three-toed sloth or any organism is taken for granted by most biologists. In fact, as evolutionists know, variability is much more interesting. Even type specimens are not taken as reliable averages or ideals of a particular species. However, for quite some time in the history of genome research there was a prevalent misunderstanding about DNA—that somehow there was something singular and essential in a DNA sequence that would make it a sort of commodity. Thus arose the great furor over the patenting and ownership of genes and even entire genomes. It turns out that being possessive about genes, especially in the legal sense, has not resulted in any obvious profits. If there were a truly unique sequence pertaining to, for example, colon cancer, the patent lawyers would have something tangible on which to litigate. However, genes are not unique sequences but a range of variations. A singular sequence submitted to NCBI and precisely recorded is but a snapshot of a moving target. A gene is a continuum of similar sequences within which we can say we have a particular range of function and beyond which we are in the territory of other genes more or less related. A gene is a concept of human construction in that its boundaries continually reflect what humans consider to be relevant and nothing more. And the situation is much fuzzier for the sequences between the genes. The good news, of course, is all of the work that needs to be done on genomic databases and the central position of computational analyses in accomplishing it.

Other challenges for the museum-like, archival databases of NCBI exist partly because of the limitations of human observers. We can readily see data diagrammed in three dimensions. With the addition of color and font style or even some type of movement we could add dimensions to our capabilities of seeing relationships. However, the intricacies of the databases at NCBI are far greater in dimension than what could be comfortably comprehended (by a human) all at once. Therefore, queries must necessarily be limited to a few parameters, and (alas) by making assumptions and choices we may fail to see other relationships as yet unimagined. As with any museum, visual presentation is a challenge and by no means solved on the scale that is required to understand genomes. Here, too, is another prominent role for computational analysis in the evolution of genomic research. Our visualization tools are simply inadequate for the scope of the problem and thus will provide many rewarding, creative opportunities for programmers specializing in that area.

Wunderkammers were often the private collections of wealthy scholars and therefore of limited access, but the sequences at NCBI are there to be used by all. Feel free to download as many files of sequences as you like. The actual design and construction of sequence databases is beyond the purview of this book. However, accessing and using individual and collections of files of sequence from databases, especially from NCBI, is exactly what this chapter is about. Instructions will be given for creating and naming files of sequence data and storing the files where your Perl program can find them. Also, you will be shown how to have your Perl program open a file, read sequence, store it in a variable, and then close the file. Furthermore in chapter 9, all of this will be done on a larger scale, as your program opens, reads, and closes one file after another from a list.

8.1 Input, Output, and Your Perl Program

Perl programs can open files of DNA sequence, read any initial header information and then the sequence, store the information in appropriate variables, and close the file when reading is completed. Because so much data is stored in so many files, understanding how to open files and read sequence and other data from files is one of the important steps toward programming in the large.

As shown in the diagram that follows, when your Perl program opens a file for reading (e.g., geneXYZ.fna), a "pipe" (formally called a file handle) connects your Perl program to the input file. Each time your Perl program reads a line of data from the file, the next line travels down the pipe to your Perl program, which can store the line of data in a variable. Alternatively, when your Perl program opens a file for writing, a pipe connects the program to an output file. Each time your Perl program writes some data from a variable in your program to the file, the data travels down the output pipe and is added to the output file.

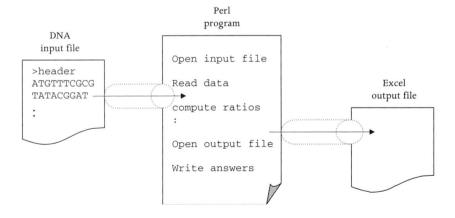

8.2 Open a File for Reading (Input)

When reading data from a file into your program, Perl's built-in open function associates the name of a data file with a user-declared file handle to move data from the file to your program. As shown earlier, it is helpful to visualize the file handle as a named pipe that connects your program to the input file. The basics are shown in the following program, which opens a file for reading, reads the initial header line from the file, and closes that file.

```perl
#!/usr/bin/perl
use strict;
use warnings;

# Drosophila melanogaster odorant receptor Or83b mRNA
my $filename = "fly_odorant_receptor.fna";

my $FNAFILE;   # create a variable to hold the file handle

# try to open the file;
# if no file by that name, print error message ($!), then exit

open ($FNAFILE, '<', $filename)
     or die "Cannot open the input file: $filename: $!";

my $header;

$header = <$FNAFILE>; # read first line of file

chomp $header;          # remove newline at end of the line

print "Header line is: \n $header \n";

close ($FNAFILE);       # close the file
```

open($FILEHANDLE, 'MODE', FILENAME **)**

The $*FILEHANDLE* is a user-defined variable, that is, you get to make up the name, for example, $FNAFILE. Because this is the name you will use in your program whenever you perform a read, we recommend that you use a good name for your file handle. The FILENAME can be one of three types of strings:

i. a simple file name (**"fly_odorant_receptor.fna"**) implying that the input file is in the same directory as the Perl program;

ii. a relative path name (**"fruit_fly/olfaction_genes/fly_odorant_receptor.fna"**) implying that the Perl program is in a directory that contains a directory called fruit_fly, which contains a directory called olfaction_genes, which contains the input file to open (fly_odorant_receptor.fna);

iii. an absolute path name, implying that the program knows the "absolute" location of the input file, for example,
Windows Users: **"C:\genes\fly_odorant_receptor.fna"**;
MacOS X and Linux Users:
"/Users/mleblanc/Desktop/fly_odorant_receptor.fna".
(If you are using TextWrangler or BBEdit editors on MacOS X, see section 8.5).

When a *FILENAME* is opened for reading, the second argument to the open function is '<'. This second argument, referred to as the MODE, means the file is to be "opened for input." A summary of various modes is shown in section 8.4. Notice the use of "or die" in conjunction with the open statement. Although not required, we recommend that you always use this safety feature when opening a file. Failure to open a file is a common occurrence for a number of reasons, including misspelling a file name. If the open function should fail for any reason, the die function will print an appropriate message. We recommend that you also print the system $! variable in this situation. This will print extra system information concerning the reason the open failed. The open statement will fail if the given file name cannot be found or the input file has permissions set to be unreadable by you.

Once opened, your program can proceed to read data from the file one line at a time by using the line input operator, <. . .>. Perl's line input operators surround your FILEHANDLE. Note that when reading from the file you use the FILEHANDLE, not the FILENAME.

```
$header = <$FNAFILE>; # read one line
```

The line input operator returns the next line in the input file or an undefined value (false) if no more data are available in the file. We will use this convention to detect the end of a file. Note that the line input operator grabs all the characters on the next line, including the end of line character (e.g., newline character). It is often the case that you do not want to store this extra character. We follow a read-line statement with Perl's chomp function that "chomps" the trailing newline character from the end of the string. If a line of input does not end with a newline character, chomp has no effect on the string; chomp only removes trailing new line characters. Finally, it is always good programming practice to close your file handle once you have finished reading from the input file.

8.2.1 Reading Lines until the End of File (EOF)

The following loop will open a file for reading and then read lines from the file until reaching the end of the file (EOF). In this example, the file is DNA coding sequence from an olfaction gene from the fruit fly genome, specifically *Drosophila melanogaster* odorant receptor AY567998.1. The first few lines of the file are shown here.

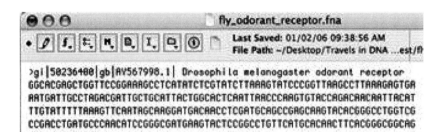

The loop to read until the end of the file (the while NOT EOF loop) will halt when an attempt to read another line (**<$FNAFILE>**) fails. The line input operator returns an undefined value, which conveniently evaluates to "false," which causes the loop to terminate.

```perl
# Drosophila melanogaster odorant receptor Or83b mRNA
my $filename = "fly_odorant_receptor.fna";

my $seq;

my $firstline;        # hold header line (ignore this line)
my $line;             # holds each subsequent line one at a time

my $FNAFILE;          # create a file handle to the opened file

open ($FNAFILE, '<', $filename)
      or die "Cannot open the input file: $filename: $!";

# read and ignore header line if found (complain if not found)
if ( !( $firstline = <$FNAFILE> ) )
{
      print "Can NOT read header line from: $filename \n";
      exit();
}

my $seq = "";

# WHILE not EOF, grab next line and concatenate sequences
while ( $line = <$FNAFILE> )
{
      chomp $line;         # remove newline from end

      $line = uc($line);   # convert to upper-case

      # concatenate new line onto end of sequence so far
      $seq = $seq . $line;

} # end WHILE not EOF

close ( $FNAFILE );

print "Sequence is: $seq \n";
```

8.2.2 Reading from a File in a Subroutine

Combining subroutines from the previous chapter and our new appreciation for opening and reading files of sequence, a subroutine to read in DNA makes a very useful utility for many of your programs. The following Perl program calls a subroutine that will open a FASTA format file and return the DNA sequence in one (long) scalar string variable.

```perl
my $mouseFilename = "Mus_musculus_chr6_clone_RP24103E20.fna";

my $mouseSequence;

# call subroutine to do all the work and return DNA sequence
$mouseSequence = readInDNA( $mouseFilename );

# . . .

#----------\
# readInDNA \
#-------------------------------------------------------------------
# SUMMARY: Subroutine to open a FASTA formatted file of DNA
# sequence and return as one long string. The function will
# remove any whitespace, ignore lines that begin with a #, and
# any digits that may appear in sequence, e.g., line numbers or
# base pair counts.
#
# IN:        1 argument: name of file holding the DNA
#            (assumed FASTA format that includes a >header line)
#
# RETURNS: DNA as one string in uppercase nucleotide letters
#
sub readInDNA
{
  my $filename = shift (@_); # save filename argument in local
                             # variable

  my $firstline;  # holds headerline (ignored)
  my $line;       # holds each subsequent line one at a time

  my $FNAFILE;    # create a file handle to the opened file

  open ( $FNAFILE, '<', $filename )
        or die "Cannot open the input file: $filename: $!";

  # read and ignore the header line if found (complain if not)
  if ( !($firstline = <$FNAFILE >) )
  {
    print "Can NOT read header line from the file called: $filename";
    exit();
  }
```

```perl
my $seq = ""; # continually concatenate each line to end of $seq
# WHILE not EOF, grab next line and concatenate sequences together
while ( $line =  <$FNAFILE>)
{
     chomp $line;                  # gobble the newline character

     # discard a blank line
     if ( $line =~ m/^\s*$/ )      # if whitespace (\s) start(^) to
                                   # end($),
     { next; }                     # return to while for next line

     # discard comment line (any line starting with a comment(#))
     if ($line =~ m/^\s*#/)
     { next; }

     $line =~ tr/0123456789//d;    # remove all digits
     $line =~ tr/ \t\n\r//d;       # remove any extra whitespace
     $line = uc($line);            # convert everything to
                                   # upper-case

     $seq = $seq . $line;          # concatenate new line onto
                                   # the entire text so far

  } # end WHILE not EOF

  close( $FNAFILE );

  return $seq;

} # end subroutine readInDNA
#-----------------------------------------------------------------
```

8.2.3 Reading Many Files One at a Time

It is often the case that you have tens (if not hundreds) of files to open and analyze. Refer to chapter 9 (section 9.4) for an elegant way to automatically save all the file names in a directory and open and process them one at time.

8.2.4 Preparing Directories of Files for Input

When you have many files to open and process, we recommend that you create separate directories to hold your input files. As mentioned, if you simply list the file name in an open function, Perl assumes that the input file can be found in the same directory as the Perl program. If you create directories and subdirectories to hold your input data, take care to use the appropriate full path name to the file(s) you want to open. For example, assume you have hundreds of files storing the downstream sequences of all the genes in many Archaea and Bacterial genomes. A directory structure to help organize these files is shown next.

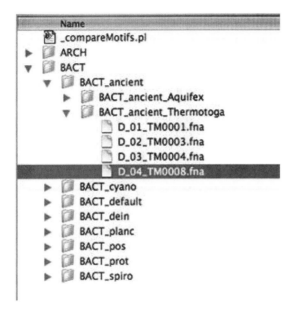

If an analysis was focused on the downstream region of a particular *Thermotoga* gene (D_04_TM0008.fna), your Perl program could concatenate a relative path name on to the front of the file name prior to the open statement, as shown below. The path name is called relative because relative to your Perl program the particular file to open can be found by looking in the BACT directory, then within that directory in the BACT_ancient directory, and so on.

```
my $pathname  = "BACT/BACT_ancient/BACT_ancient_Thermotoga/";

my $filename  = "D_04_TM0008.fna";

my $fullname = $pathname . $filename;

open ($FNAFILE, '<', $fullname)
       or die "Cannot open the input file: $fullname: $!";
```

Section 9.4 will show you how you can automate this idea for all the files in a particular directory.

8.3 Opening a File for Writing (Output)

It may soon become apparent that your Perl programs are able to produce more output than you can or want to manage on your console. Moreover, you may want to conduct further analysis on your output, or your group may need the final output to be in a more reader-friendly format than Perl can provide, including graphs. Perl is great at manipulating strings and files of sequence, but this doesn't mean that it should do everything. In regard to output, we recommend that you use Perl to read your input sequence files, perform your analyses, and then print the tab-delimited or comma-delimited results to an output file that can be opened for later use by another tool, for example, Excel, Maple, or R.

To open a file for output, we use the output mode, '>'. Unlike opening a file for reading, the '>' mode will either (1) overwrite the contents of an existing file if a file with that name already exists, or (2) create a new file with that name. Clearly, the first case can be potentially dangerous and you should be careful to avoid reusing a file name that you do not want overwritten.

```perl
#!/usr/bin/perl
use strict;
use warnings;

my $outputFilename;

#-----------------------------------------------------------
# OPEN the OUTPUT file (a tab-delimited file for Excel)

my $OUTPUT_HANDLE;

$outputFilename = "output.xls";

open ( $OUTPUT_HANDLE, '>', $outputFilename )
        or die "Cannot open the file: $outputFilename: $!";
#-----------------------------------------------------------
my $numA = 135;
my $numT = 158;
my $total = 612;

my $perA = ($numA / $total)*100;
my $perT = ($numT / $total)*100;

printf $OUTPUT_HANDLE "Percent A\tPercent T \n";

printf $OUTPUT_HANDLE "%6.2f\t%6.2f", $perA, $perT;

close ( $OUTPUT_HANDLE );
```

Before your run this program, the directory OutputTest contains only your Perl program.

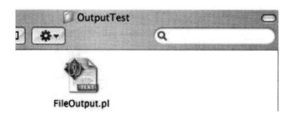

FileOutput.pl

After you run the program, the directory will contain your new output file, `output.xls`.

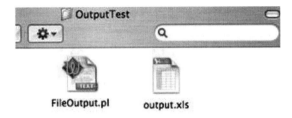

If you double-click on the `output.xls` file, Excel will open it, and your tab-delimited output will appear in the expected column format as shown here.

	A	B	C	D
1	Percent A	Percent T		
2	22.06	25.82		
3				
4				

Box 8.1 Ten Ways to Look at DNA

Immersed as we are in the analysis of sequences and about to make a foray into the dense files of NCBI, we sometimes can forget that DNA is foremost a molecule, an extraordinarily long one, with peculiar slime-like properties. To help you to keep this in mind as you labor over As, Cs, Gs, and Ts, here are 10 short projects or lab exercises, some of which are simple enough to do at home or outdoors or in the bathroom down the hall from the computer lab. The full instructions for some projects are in the Appendix.

1. Isolate DNA from strawberries (see Appendix for instructions).
2. Build a DNA molecule from a kit or with Legos. For example, Lego sculptor Eric Harshbarger, who is also a Perl programmer, made a Lego model of the double helix, the approximate instructions (or suggestions) for which are found online at http://ericharshbarger.org. Another set of instructions for Lego DNA appears at http://ncbe.reading.ac.uk/DNA50/lego.html (reprinted from March 2003 *School Science Review*). The site also maintains an astonishing variety of DNA celebratory items.
3. Reading about DNA as a topic of science fiction can be frustrating if the author does not have all of the details right. Instead of making specific

continued

Box 8.1 Continued

recommendations in that genre, we suggest reading up on linguistics and mentally inserting the word *DNA* in the examples. We have noticed that Steven Pinker's *Words and Rules* works well for this exercise, as well as being fascinating on the topic of linguistics.

4. If you or a family member has a karyotype done at a hospital, request that you be able to keep a copy of the chromosome photographs. That would include a karyotype from an amniocentesis that is a sort of first photograph.

5. Look at domesticated plants as homeotic (gene regulatory) mutants. For example, the diversity of brassicas is easily demonstrated at a grocery store with cabbage, brussel sprouts, cauliflower, and broccoli. The same can be done with the extremes of domestic dog breeds.

6. Look at variegation in plants as evidence of jumping gene or viral activity, signs of the dynamic genome. For example, some ornamental plants are bred for the trait of variegation. The more splotchy and irregular and unstable the pigmentation is, the more likely it due to unstable genetic elements. Tulips with "broken colors" are a famous example that in the seventeeth century caused the furor of Tulipmania.

7. Look at variability in either domestic or wild animals and plants as manifestations of alleles. Pigmentation is an obvious choice, for example in purple/red versus white eggplant, cabbage, lettuce, and other domestic plants. The pigment is anthocyanin and the simplified view of the two alleles is A_ makes anthocyanin and aa does not.

8. Try a sequence analysis on a small scale by hand using colored pencils or some other artistic medium. This is recommended in general as a first step to thinking about an algorithm for sequence analysis.

9. DNA (and its associated historical milestones) can be a reason for celebration. The birthday of the publications of the structure of DNA is April 27, 1953. Festive items might include special T-shirts, mugs, pencils, and so on with DNA sequences printed on them.

10. Enjoy some of the beautiful graphic representations of DNA and its functions, including models, paintings, and especially Web sites and DVDs with action sequences.

8.4 A Summary of Modes for Opening Files for Input and Output

The following table summarizes the most common modes that are used when using Perl's open statement.

Mode	Can read from	Can write to	Can append to end	Create new file if not there	Overwrite existing file
<	Yes	No	No	No	No
>	No	Yes	No	Yes	Yes
>>	No	Yes	Yes	Yes	No

8.5 Opening Files on the Mac When Using TextWrangler or BBEdit

If you are using a Macintosh (OS X and higher) *and* a text editor such as TextWrangler or BBEdit, these editors will not recognize your relative path names, thus you'll get error messages when attempting to open your files. You can ignore this section if you are using a Windows operating system or if you are running your Perl programs on a Linux or Mac OS X command line.

Because TextWrangler and BBEdit are text editors and not full-fledged Perl development environments, they are not able to recognize by default the directory where your Perl program runs. The following solution will help. It contains some features of Perl that have not yet been addressed, but consider this an exercise in borrowing code. In short, this solution borrows functionality from a module of Perl functions called File::Spec. In particular, the splitdir and catfile functions help determine the absolute path name of your running Perl program, thereby helping you construct the appropriate path name to use to open your files.

```perl
#!/usr/bin/perl
use strict;
use warnings;

# needed for MacOS X if using BBedit or TextWrangler
use File::Spec;

#-------------------------------------------------------------
# establish the absolute pathname to directory of input files
#-------------------------------------------------------------
# need for MacOS X when using BBedit or TextWrangler
#
# determine the absolute pathname to the current directory
my @path = File::Spec->splitdir( File::Spec->rel2abs($0) );

my $abs_pathname;
$abs_pathname = File::Spec->catfile(@path[0..$#path - 1], "");

my $pathname  = "BACT/BACT_ancient/BACT_ancient_Thermotoga/";
my $filename  = "D_04_TM0008.fna";

my $fullname = $abs_pathname . $pathname . $filename;

my $FNAFILE;

open ($FNAFILE, '<', $fullname)
      or die "Cannot open the input file: $fullname: $!";

print "full pathname opened is: \n $fullname \n";

#...

close ( $FNAFILE );
```

The output of the automatically determined absolute path name is shown below:

```
full pathname opened is:
/Users/mleblanc/InputTest/BACT/BACT_ancient/BACT_ancient_
Thermotoga/D_04_TM0008.fna
```

The same procedure is needed when you want to open a file for output. `File::Spec`, `splitdir`, and `catfile` can be used to build an absolute path name. This can be concatenated with your output file name and opened for writing.

8.6 Finding Files of Entire Genomes at NCBI

NCBI is a rich source of genomic data. One of our favorite sections is the collection of prokaryotic genomes; this serves as a good first source of data. Although your colleagues will undoubtedly have suggestions for appropriate data sets, if you are looking for places to start, we recommend you pick a microbial genome (e.g., *E. coli* or some other exotic microbe you've heard mentioned) and download at least two files: the DNA (stored in a FASTA formatted file of 70 nucleotides per line followed by a newline; the file name ends with the extension .fna) and the associated protein table (file extension .ptt), which includes annotation on the protein coding regions (genes) in this organism's genome, including but not limited to the starting and stopping base pair locations of the genes, the accession (ID) numbers for each gene, and possibly a description of this gene's function.

The following (ftp) URL will take you to the collection of microbial genomes:

ftp://ftp.ncbi.nih.gov/genomes/Bacteria

The contents of this page provides links to hundreds of genomes, the first few of which are shown here.

Bacteria

Anonymous access granted, restrictions apply.

Path: [ftp.ncbi.nih.gov][genomes][Bacteria]

Name	*Kind*	*Last Modified*
1	Document	Mon, Mar 7, 2005, 12:00 AM
Acinetobacter_sp_ADP1	Folder	Thu, Aug 18, 2005, 3:54 PM
Aeropyrum_pernix	Folder	Thu, Aug 18, 2005, 3:54 PM
Agrobacterium_tumefaciens_C58_Cereon	Folder	Thu, Aug 18, 2005, 3:54 PM
Agrobacterium_tumefaciens_C58_UWash	Folder	Thu, Aug 18, 2005, 3:54 PM
Anabaena_variabilis_ATCC_29413	Folder	Wed, Sep 21, 2005, 4:24 PM
Anaplasma_marginale_St_Maries	Folder	Thu, Oct 27, 2005, 4:53 AM
Aquifex_aeolicus	Folder	Thu, Aug 18, 2005, 3:54 PM

Archaeoglobus_fulgidus	Folder	Thu, Aug 18, 2005, 3:54 PM
Azoarcus_sp_EbN1	Folder	Tue, Oct 4, 2005, 10:50 PM
Bacillus_anthracis_Ames	Folder	Thu, Aug 18, 2005, 3:54
:		
:		

The page for Acinetobacter_sp_ADP1 is shown next. Note the .fna and .ptt files. Click on these two files to download the genome and protein table files, respectively. Once downloaded, we recommend that you rename these files to include the genus and species names, for example, rename NC_005966.fna to Acinetobacter_sp_ADP1.fna.

ftp://ftp.ncbi.nih.gov/genomes/Bacteria/Acinetobacter_sp_ADP1

Acinetobacter_sp_ADP1

Anonymous access granted, restrictions apply.

Path: [ftp.ncbi.nih.gov][genomes][Bacteria][Acinetobacter_sp_ADP1]

Name	Size	Kind	Last Modified
NC_005966.GeneMark	799K	Document	Mon, Dec 5, 2005, 2:55 PM
NC_005966.GeneMarkHMM	197K	Document	Mon, Dec 5, 2005, 2:55 PM
NC_005966.Glimmer2	156K	Document	Mon, Dec 5, 2005, 2:55 PM
NC_005966.asn	10,881K	Document	Sun, Dec 4, 2005, 2:57 PM
NC_005966.faa	1,364K	Document	Sun, Dec 4, 2005, 2:57 PM
NC_005966.ffn	3,245K	Document	Sun, Dec 4, 2005, 2:57 PM
NC_005966.fna	3,564K	Document	Sun, Dec 4, 2005, 2:57 PM
NC_005966.frn	44K	Document	Sun, Dec 4, 2005, 2:57 PM
NC_005966.gbk	8,701K	Document	Sun, Dec 4, 2005, 2:57 PM
NC_005966.gff	4,870K	Document	Sun, Dec 4, 2005, 2:57 PM
NC_005966.ptt	291K	Document	Sun, Dec 4, 2005, 2:57 PM
NC_005966.rnt	6K	Document	Sun, Dec 4, 2005, 2:57 PM
NC_005966.rpt	234 bytes	Document	Sun, Dec 4, 2005, 2:57 PM
NC_005966.val	4,990K	Document	Sun, Dec 4, 2005, 2:57 PM

Open the files with a standard editor and consider their contents. Aren't they beautiful?

....

Going Back for More

1. Follow the directions in section 8.6 and download a file of DNA sequence and the associated protein table file. Use these files in the following exercises.
2. Open a file for reading and print the first line in the input file such as shown in section 8.2.
3. Cause the open function to fail by changing the file name in your program to the wrong file name. Carefully read the error message so you will recognize this type of message in the future.

4. Open a file for reading and another file for writing. Read each line from the first file and output the lines to the second file. In effect, you are writing the software that handles the Copy File or Duplicate File utility that is available in your operating system.

5. Open a FASTA format file of DNA sequence for reading and read each line in the file. For each line, calculate the A, C, G, and T percentages and output the results to a new tab-delimited output file (e.g., output.xls). After you run the Perl program, open your output file with Excel and create a bar graph of your results.

6. Modify the `readInDNA` subroutine shown in section 8.2.2 to open a file of protein sequence (rather than DNA). The subroutine should open, read, and return all the lines of protein sequence in one scalar string variable.

9 Arrays

In which a list-like data structure is employed to collect and manage large numbers of DNA files and sequences for analysis.

Now the calculating machine . . . does—being moved by the turning of a crank—calculate . . . tables necessary to the labors of all devoted to the higher branches of mathematics. Feed it with figures, set it properly, and it will turn out the required results with unfailing accuracy and with great speed.

—Charles Babbage, quoted in Harper's New Monthly Magazine
"The Life of a Philosopher" (1867)

This chapter is about scaling up. Any analysis that can be done on a single string can be done repeatedly on many strings. Any analysis that can be done on the sequence in one file can be done on the sequences from many files, one after another. An array is a list-like data structure that allows you to organize your sequence analyses on a grander scale. Arrays can hold almost any set of items and allow you to access and manipulate them with various functions. Note that all of the items stored in arrays are essentially scalar variables of the sort discussed in chapter 3. The following are items (scalar variables) that biologists especially may wish to organize in the accessible list-like fashion of an array. A list of short motifs:

```
"agga", "caat", "cggc", "cttc", . . . ,
```

a list of gene sequences:

```
"atg . . . . . . tag", "atg . . . . . . taa", . . . ,
```

or a list of file names containing coding sequence:

```
"NM_012374_1.fna", "XM_523807.fna", . . . .
```

One of the great advantages of being able to list files by name is being able to request that those files be opened up one by one, the sequences within examined and analyzed, and then the file closed and the next one opened, again and again until all are done. The management and use of huge numbers of files is a critically important task in bioinformatics, so this chapter will linger on some of the details of dealing with large numbers of idiosyncratic file names. There will also be some suggestions for how arrays can help you parse useful information from lengthy strings. For example, you might like to capture and organize some or all of the details in the string:

```
"Escherichia coli K12 at Univ. Wisconsin [Complete]"
```

such as genus name, species name, strain number, and whether the genome sequencing is complete or in progress.

9.1 A Quick Example: An Array of Motifs

Assume you have hundreds of putative regulatory motifs that you need to store. An array can store these motifs for you. The following Perl snippet fills an array, @motifs, with the first five of those motifs.

```
# an array to store many putative regulatory motifs

my @motifs = ( "ACGT", "CAAT", "TATATCTT", "GCAT", "CGCGC" );
```

The array @motifs holding five scalar strings would appear like this in memory, each string stored within one cell of the array. Each cell of the array has an index (or subscript) with indices numbered beginning with zero.

"ACGT"	"CAAT"	"TATATCTT"	"GCAT"	"CGCGC"
0	1	2	3	4

Syntactically, notice that when you refer to an entire array, the array name is prefaced by the @ symbol. To refer to one specific scalar element within the array, you need to "index the array," for example, $motifs[3] would access the array element subscripted by a 3 and get you the motif GCAT.

In addition to accessing individual elements in the array, Perl provides a number of looping control structures to help you visit every element in the array, one at a time. For example, the following foreach loop can visit every element of the array, one at a time for a set of analyses of your choosing. For now, suppose we would like to print any of the motifs that are longer than seven (7 bp) and/or GC-rich (GC content only).

```
my @motifs = ( "ACGT", "CAAT", "TATATCTT", "GCAT", "CGCGC" );

# for each of the motifs in the array (one at a time),
# is the next motif longer than 7 bp or GC-rich?

foreach my $nextMotif (@motifs)
{
    if ( (length($nextMotif) > 7) or ($nextMotif =~ m/^[GC]+$/) )
    {
        print "$nextMotif \n";
    }
}
```

The output is shown below; the third motif is longer than 7 bp and the last motif is GC-rich.

```
TATATCTT
CGCGC
```

The powerful part of this example is the array. If the array held hundreds of motifs, you would not have to change any of the code in the `foreach` loop. The remainder of this chapter will show you how to fill the cells of an array with various types of information and then process some or all of those cells.

9.2 **Arrays as a Data Structure**

The array is one of Perl's two aggregate data structures. The other aggregate data type, the hash table, is introduced in the next chapter. Perl has three built-in data structures: (1) *scalars*, (2) *arrays* of scalars, and (3) *hashes* of scalars (sometimes referred to as *associative arrays* of scalars, although we will stick with hash or hash table). The scalar is a fundamental data structure (a variable), most often storing a simple string (ACGTACGT) or numerical value (13 or 0.25), whereas arrays and hashes are larger, more complicated data structures that store aggregates of scalars. Arrays are an ordered list of scalars with each item in the list indexed or referenced by an integer subscript. Pictures of a single scalar and an array of many scalars can begin to accent the utility of storing many like-scalars all together in one place.

```
my $oneFileName; # scalar variable to hold one(1) filename

$oneFileName = "NM_012374_1.fna";
```

```
"NM_012374_1.fna"
```

$oneFileName

```
my @allFiles; # array to hold many filenames for an experiment
  :
# assume the initial four cells in the array get filled here...
```

@allFiles

"NM_012374_1.fna"	"XM_523807.fna"	"AL606805_21.fna"	"AY448876_2.fna"
0	1	2	3

9.3 Motivation for Using Arrays

Suppose your group needed to run some experiments on prokaryotic genomes. Following good software engineering advice, you decide to start with just three genomes to ensure that you can successfully open the genome data files and check that your logic and math are correct. You might store the names of the files containing the FASTA format genome data in one scalar variable per genome.

```
my $genome1 = "Bacillus_anthracis_str_Ames.fna";
my $genome2 = "Escherichia_coli_K12.fna";
my $genome3 = "Pyrococcus_furiosus_DSM_3638.fna";
```

Assuming that your experimental tests on these three genomes go well, your group leader peers over your shoulder and says, "Great. Now can you make that work for the 200+ bacterial genomes available at NCBI and why not also include the scores of Archaea genomes as well?"

As you look at the Perl, code, you suddenly get this awful image in your head of the following:

```
my $genome001 = "Acinetobacter_sp_ADP1.fna";
       :
my $genome013 = "Bacillus_anthracis_str_Ames.fna";
       :
my $genome102 = "Escherichia_coli_K12.fna";
       :
my $genome179 = "Pyrococcus_furiosus_DSM_3638.fna";
       :
my $genome214 = "Zymomonas_mobilis_subsp_mobilis_ZM4.fna";
```

Clearly, this is a programming nightmare. Even if you were to get this program to work with all these independent variables, what would you do if your group decides to remove all the *Gammaproteobacteria* genomes? This is a classic case of a program that works *versus* a program that has been designed to handle the general case. In this example, we need a program that works for *any* number of genome files and, more specifically, a program that can store any given number of file names in a data structure without having to change the program to accommodate the varying number of files on a particular run.

The take-home story from this example is that scalar variables alone are not sufficient to store a list of many related items, especially when you need to write a program that can operate on those items one at a time. The next section reveals a solution for holding multiple file names by using an array.

9.4 Storing "a Glob" of File Names

A good example of using an array is to store an unknown number of file names that are stored in a directory (or folder) into an array. Perl's **glob** function will peek into a given directory and store the name of each file into an array, one per cell in the array. Assume there is a directory called bacteriaGenomes holding the DNA sequence data for many complete genomes.

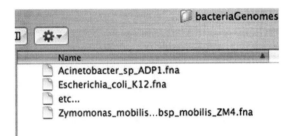

We can use glob to load all the file names found in this directory into an array.

```
my @bacteriaFilenames; # an array of many bacterial filenames

# a directory called "bacteriaGenomes" holds the files

# the /* means to get all the files in that directory

my $pathName = "bacteriaGenomes/*";

# grab a "glob" of files, each of the filenames
#       is put into individual elements of the array

@bacteriaFilenames = glob("$pathName");
```

Once glob has done its work, the array @bacteriaFilenames would be filled with all the file names in the directory bacteriaGenomes, with each cell of the array filled with one name. The diagram below shows the initial two cells of the array.

"…/Acinetobacter_sp_ADP1.fna"	"…/Escherichia_coli_K12.fna"	
0	1	2

Notice that each cell in an array is numbered and the numbering begins with zero. These numbers are referred to as subscripts or indices. Array subscripts provide immediate access to individual cells in the array. We return to array subscripts in the next section.

glob(*EXPRESSION*)

The glob function returns a list of file names. The argument for glob is a file path identifier (*EXPRESSION*), which specifies how to find the files and which file names to match and return. File name expansions occur such as you might encounter on a Linux or MS-DOS command line, for example, a * means match any sequence of characters including none. Note glob file name syntax is close to but not the same as the full suite of regular expression syntax. The following examples show potential uses of glob.

```
my @filenames; # an array to hold filenames
# glob all files (*) in the current directory
# that end with the extension ".fna"
@filenames = glob("*.fna");

# glob all files (*) in the directory Archaea/
# that end with the extension ".fna"
@filenames = glob("Archaea/*.fna");

# glob all files in the directory Bacteria/ProteinTables/
# that begin with an 'E' and end with the extension ".ptt"
@filenames = glob("Bacteria/ProteinTables/E*.ptt");
```

Special Cases

Two important special cases are worth mentioning.

1. A file name that begins with a dot (.) is ignored by glob unless you've specifically requested that this character be explicitly matched in the *EXPRESSION*. In Linux, this means that glob will not return the current (.) directory nor the up one level (..) directory.
2. glob may *not* work on file names that contain whitespace. Thus, a file name containing a blank space, for example, Ecoli K12.fna, may not be handled by glob. Names that contain spaces should either be renamed, for example, by replacing the blank with an underscore Ecoli_K12.fna, or if you are faced with too many names that contain whitespace, you'll need to use the Perl module File::Glob, which handles whitespace in file names (see chapter 14).

Try This Now

The ability to glob files and deal with them one at a time is a must-have tool available to you. Whether you need to peek at the header line of hundreds of files of coding sequence or you are running an experiment on scores of genomes, the glob function and foreach loop encourage you to organize your input (sequence data) files in directories, glob the file names into an array, and then operate on those files one at a time.

Create a new Perl program (e.g., globTest.pl). In the same directory as your Perl program, create a new directory called Tests. In the Tests directory, create and save four or five empty files (e.g., one.txt, two.txt, etc.). In this example, the files need not contain any data because we'll just attempt to glob the file names and print them. Write a short Perl program to glob the files in the directory Tests and print the names to the console.

9.5 Iterating through an Array with `foreach`

Once an array is populated with values such as provided by glob, Perl's **foreach** control structure will select each of the items in the array one at a time to allow you to process each of the items individually. The foreach statement is a control structure similar to the while loop introduced in chapter 6, except in this case the foreach loop executes once for each item in an array.

Assume you want to run an experiment on many files of DNA sequence. The following lines of Perl will glob all the file names in the directory bacteriaGenomes into an array and then iterate through each of those files in the array one at a time using a foreach loop. "For each" of the file names in the array, the next file name in the array is stored in the variable $nextFile, and that file is opened and processed depending on your experiment. Once control reaches the bottom of the foreach loop, control returns to the top of the foreach loop, where the next file name is retrieved from the array, stored in the variable $nextFile, and the process repeats.

```
my @bacteriaFilenames; # an array of filenames
my $nextFile;           # hold one filename at a time

@bacteriaFilenames = glob("BacteriaGenomes/*");
```

".../Acinetobacter_sp_ADP1.fna"	".../Escherichia_coli_K12.fna"	
0	1	2

```
# now the array @bacteriaFilenames holds the filenames

# for each filename in the array, process the filename

foreach $nextFile (@bacteriaFilenames)
{
        my $sequence;
        $sequence = readInDNA($nextFile);
                :
                :
} # end foreach loop (return to top)
```

Box 9.1 Favorite Gene (Annotation at the Grass roots)

Scaled-up, public, and private gene-finding programs in particular and genome analysis in general provide much of the annotation at databases such as those found at NCBI. Many genome search projects are motivated by commercial or political interests. However, details and nuances are easily missed by comprehensive programming approaches. The sheer volume of data precludes lingering and reflection. That is the reason, in many cases, that the annotation and analysis of genes (and proteins) and their associated promoter sequences are best done on a grassroots level by the very labs that already have a focus and vested interest in them, such as your lab.

A logical place to start in bioinformatics is with your own favorite genes. (We call them *favGenes* after a program we wrote to find them quickly in large databases.) Begin by taking ownership of your sequences by going to NCBI and downloading the files (see chapter 8) and then do what the big commercial programs cannot afford to do: explore the sequences, motivated by nothing more than curiosity. Who better to do that but you? A sequence annotation tells a story of its own about the knowledge base, focus, and priorities of the annotator.

Consider printing out your favorite sequences in some linear format that can be taped up along the wall of a long corridor or around the walls of your lab. Get to work with colored markers to do some hand annotation, such as marking the beginnings and ends of the genes. In a way, you are creating your own meta-version of the sequence with layers of new potential information, perhaps noticed for the first time by you. Obviously, you should do some analyses and annotations with the wonderful public software available at NCBI, such as BLAST. However, by learning some Perl, you have the opportunity to take the analyses much further. Design and implement your own dream software for searching the kinds of motifs, patterns, and associations that interest you.

9.6 Determining the Number of Elements in an Array

Once an array is loaded with data, you will often need to know how many individual elements are held in it. For example, if you glob a directory of files into an array, you'll probably want to report the number of files in your experiment. A scalar evaluation of the name of an array will result in its length, or the number of elements in the array.

```
my @bacteriaFilenames; # an array of filenames

@bacteriaFilenames = glob("BacteriaGenomes/*.fna");

my $lengthOfArray = scalar(@bacteriaFilenames);

print "BacteriaGenomes/ holds $lengthOfArray .fna files";
```

Note that because array subscripts begin with zero, the length of an array is always one larger than the subscript of the last cell in the array that is filled with data. The exact value of the last subscript of an array can be obtained with $#*arrayName*, for example, **$#bacteriaFilenames** evaluates to the value of the subscript of the last element in the array @bacteriaFilenames. For arrays that contain data, the last subscript in an array is always one less than the length of the array; that is, the following is always true:

```
scalar(@arrayName) == ( $#arrayName + 1 )
```

The following example manually fills a small array and then shows an example of determining the length and ending subscript of an array.

```
my @patronSaints; # an array of names of our "patron saints"
@patronSaints = ("Chargaff", "Franklin", "Kent", "Wall");
```

"Chargaff"	"Franklin"	"Kent"	"Wall"
0	1	2	3

```
my $lengthOfArray = scalar(@patronSaints);

my $lastSubscript = $#patronSaints;

print "The length of the array is: $lengthOfArray \n\n";
print "The last subscript in the array is: $lastSubscript \n";
```

```
The length of the array is:  4
The last subscript in the array is:  3
```

9.7 Working with Arrays Using Subscripts

In addition to using Perl's foreach loop for iterating through all the elements in an array, you will often need to access individual elements, iterate through only some of the elements by using the subscript (index) of individual cells, or take a slice of an array such as the third, fourth, and fifth elements.

For example, assume an array @bacteriaFilenames is filled with file names.

```
my @bacteriaFilenames;        # an array of filenames

@bacteriaFilenames = glob("BacteriaGenomes/*.fna");
```

".../Acinetobacter_sp_ADP1.fna"	".../Escherichia_coli_K12.fna"	
0	1	2

To reference an individual cell in the array, the array name is prefaced with $ and is subscripted with the subscript of interest, for example, to access the initial element in the array:

$bacteriaFilenames[0] *means the scalar value in cell #0*

Using the example of file names above, this reference to the array at subscript zero evaluates to the string ending with the name /Acinetobacter_sp_ADP1.fna because that is the string value stored in the initial (zeroth) cell of the array.

$bacteriaFilenames[1] *means the scalar value in cell #1*

This reference evaluates to the string ending with /Escherichia_coli_K12.fna, because that is the string value stored in the array element indexed with the subscript [1].

Because the subscripts are consecutive integers, you can use a loop in conjunction with a variable that keeps track of the current cell. More specifically, if you use the variable $i to index your array and the variable $i holds the value zero (0), then the reference $bacteriaFilenames[$i] refers to the zeroth element in the array. If you add one to $i so that the variable $i holds the value one (1), then the reference $bacteriaFilenames[$i] refers to the next element in the array, and so on.

The following while loop uses $i to index the elements of the array one at a time.

```
@bacteriaFilenames = glob("$pathName");

# determine the number of filenames in the array
my $sizeOfarray = scalar(@bacteriaFilenames);

my $i = 0;
while ($i < $sizeOfarray)
{
        print "File #[$i] is: $bacteriaFilenames[ $i ] \n";

        $i = $i + 1; # move down to the next element

} # end while

print "That's all of them.\n";
```

9.7.1 **foreach** Loop: Working on Parts of Arrays

In classic Perl fashion, there are many ways to iterate through some or all of the elements in an array. As we have seen, the foreach loop visits all the elements in an array one at a time from left to right. In addition, the while loop with an associated index variable (e.g., $i) that is incremented each time through the loop can also be used to iterate through some or all of the elements. Perl also allows you to use the foreach loop to iterate through a list of index values.

```
my @patronSaints; # an array of names of our "patron saints"

@patronSaints = ("Chargaff", "Franklin", "Kent", "Wall");

# for each of the array subscripts, zero to last subscript
foreach my $i (0..$#patronSaints)
{
        print "Patron Saint ($i): $patronSaints[ $i ] \n";

} # end foreach
```

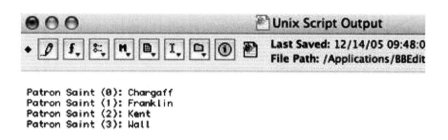

```
Patron Saint (0): Chargaff
Patron Saint (1): Franklin
Patron Saint (2): Kent
Patron Saint (3): Wall
```

But what if you do not want to process *all* the elements in the array? For example, what if you only need to process the elements in the array from index 20 to 29? The foreach loop and range operator (..) allow you to do this.

```
process array elements with subscripts [20 <= $i <= 29]
foreach $i (20..29)
{
...someArray[ $i ]...

} # end for
```

9.7.2 Taking Slices of Arrays

Perl's range operator (. .) enables you to take slices of arrays. Intuitively, a slice of an array returns a portion of it, where the portion returned is a list of values. Because a slice returns a list of values, presumably more than one, the left-hand side of your assignment statement must be ready to store multiple values. In Perl, when parentheses surround a group of variables, the group is referred to as a list, such as:

```
($second, $third)
```

When used on the left-hand side of an assignment operator (=), the variables in the list are assigned the items from the list on the right-hand side, for example, a slice of two values from an array such as shown here.

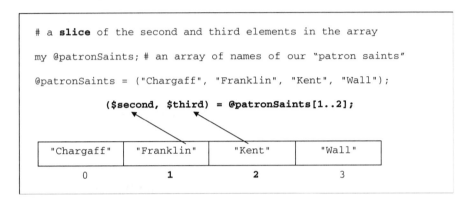

Slicing into a small number of variables works fine; however, if you need to take a large slice from an array, setting up individual variables to catch each of the items from the slice is prohibitive. Similar to our original motivation for needing arrays, we can use an array on the left-hand side of the assignment to catch a large slice from an array. The following lines of Perl store a slice of the elements from subscript 10 to 24 from one array into another array using only one line of code.

```
# store a slice of one array into another array
my @smallArray = @bigArray[10..24];
```

9.7.3 Accessing Every *n*th Element in an Array

If you only needed to process every *n*th element in an array (for example, every subscript that is odd or every third, fourth, fifth, or *n*th element), you need to return to the while loop. For example, suppose you need to process every third element starting at subscript 10 through 50. The foreach loop assumes an integer +1 increase between each element in a list of values to try (e.g., 5..10 means 5, 6, 7, 8, 9, and 10), so it will not enable you to efficiently jump to every third element. On the other hand, a while loop requires you to control the increment of your index variable, so a jump of three ($i + 3) is easily handled, as shown next.

```
# process every third element from subscript 10 to 50

my $i = 10;
while ($i <= 50)
{
        ...someArray[ $i ]...;

        $i = $i + 3; # move down to every third element

} # end while
```

9.8 Using **split**

Now that arrays are available, we introduce two additional built-in string functions, split and join. The split function "splits" a large string into a set of substrings to enable each substring to be stored in a cell of an array. Inversely, the join function introduced in the next section "joins" all the individual elements stored within the cells of an array into a single item.

split(/PATTERN/, STRING, [LIMIT] **)**

The split function repeatedly searches a STRING for characters acting as separators as defined by a regular expression PATTERN and splits the STRING into a list of substrings. The split function returns a list of substrings. Because the PATTERN is a regular expression, the separators may not necessarily be the same nor of the same size. If the PATTERN finds a match only once within the larger STRING, then two substrings would be returned; if two separator matches are found, three substrings would be returned, and so on. If no separator match is found, the entire original STRING is returned. Instead of inserting a PATTERN, it is possible to split on anything with split (//, STRING, [LIMIT]); this would allow you to break a string into individual letters.

As indicated by the brackets, the use of the third argument, LIMIT, is optional. If LIMIT is specified and positive, then split returns only up to LIMIT items, assuming the STRING contains that many delimiters. A negative LIMIT is treated as a very large positive value (don't do that). Extra variables on the left-hand side of split that do not receive values are obviously undefined, thus, you should take care to check for and handle unusual situations. When regular expressions are involved, careful programmers expect the unexpected.

9.8.1 Parsing Genus, Species, and Strain Names

Assume we are reading from a file where each line lists a number of items about *E. coli*, including the full genus and species name, the strain, where this strain was sequenced, and the status of the sequencing project. We want to access and enumerate only the strain names. As of this writing, according to NCBI there are

15 strains of *E. coli* that are fully or partially sequenced. The first six lines of the file are shown here.

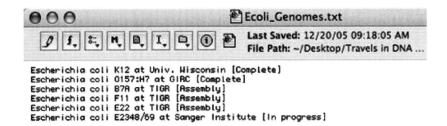

```
Escherichia coli K12 at Univ. Wisconsin [Complete]
Escherichia coli O157:H7 at GIRC [Complete]
Escherichia coli B7A at TIGR [Assembly]
Escherichia coli F11 at TIGR [Assembly]
Escherichia coli E22 at TIGR [Assembly]
Escherichia coli E2348/69 at Sanger Institute [In progress]
```

Notice that the initial three items (genus, species, strain) on each line are separated by whitespace (blank spaces in this case). Perl's split function can split items that are separated by whitespace by splitting on the pattern that contains one space " ". Note that this pattern matches any sequence of whitespace, for example, one space or multiple spaces or one tab or multiple tabs. Splitting on " " is a special case of split. In the Perl code that follows, we open the file, read each line until the end of the file, split each line into the first three items, and enumerate the strain names. We split only on the first three values because this is all we need at this point. Note also that we must capture the first two words even though we only want the third item.

```perl
# print STRAINs of E.coli

my $filename = "Ecoli_Genomes.txt";

open (INPUT, $filename)
        or die "Cannot open file: $filename --$!";

my $genus;
my $species;
my $strain;

my $nextLine;
my $i = 1;

print "Strains of E.coli fully or partially sequenced\n";
while ( $nextLine = <INPUT> )
{
        # split terms by whitespace
        ($genus, $species, $strain) = split( " ", $nextLine );
        print "[$i]: $strain \n";
        $i++; # same as $i = $i + 1
}
close INPUT;
```

The output of this code segment is shown below.

```
Strains of E.coli fully or partially sequenced
[1]: K12
[2]: O157:H7
[3]: B7A
[4]: F11
[5]: E22
[6]: E2348/69
```

9.8.2 Parsing Lineage with Split and Arrays

This example highlights a case where we are unsure of the number of items to split. In this case, the results of the split in list context can be assigned to an array.

```perl
# split lineage and print

my $pyro = 'Archaea; Euryarchaeota; Thermococci; Thermococcales; '
          . 'Thermococcaceae; Pyrococcus; Pyrococcus furiosus DSM
          3638';

my @lineage = split( /;/, $pyro);

for my $next (0..$#lineage)
{
        # continually indent deeper and deeper
        for my $indent (0..$next) { print " "; }

        print "$lineage[$next] \n";
}
```

The output is shown next.

```
Archaea
    Euryarchaeota
        Thermococci
            Thermococcales
                Thermococcaceae
                    Pyrococcus
                        Pyrococcus furiosus DSM 3638
```

9.8.3 Parsing FASTA Format Header Lines

DNA sequence files in FASTA format begin with a header line, an initial line providing information about the sequence found in the file. The header line often begins with a greater than character (>), and the individual pieces of information in the header line are typically separated by a delimiter character, often a vertical bar | (sometimes called a pipe). A sample header line is delimited by pipe characters (|):

```
>gi|49175990|ref|NC_000913.2| Escherichia coli K12, genome
```

The regular expression in the split function can match each pipe character, access the individual substrings of information between the pipes, and return the substrings in a list context to be stored individually, as shown in the following example.

```perl
my $header = ' >gi|49175990|ref|NC_000913.2| Escherichia coli
          K12, complete genome';

chomp($header); # remove newline if any

my @headerFields = split(/\|/, $header);

my $NCBI_number = $headerFields[ 3 ];
my $description = $headerFields[ 4 ];

print "The fourth item in the header is: $NCBI_number \n";
print "The fifth item in the header is: $description \n\n";
```

Note that the pattern in the regular expression contains the pipe character (|), but because this character is part of the regular expression syntax (to allow multiple possible matches, match this "or" that), we must precede the pipe character with a backslash (\). Thus, the combination of (\|) means that we really mean that the pipe character should be matched as the delimiter or in other words, do *not* (\ ) treat the | as a regular expression or-symbol.

9.8.4 Converting a String into an Array of Characters Using Split

The split function makes it easy to break a DNA sequence into individual nucleotides or a protein sequence into individual amino acid symbols and store each individual item into an array. In this case, you split on "nothing" (//), this being a default way to divide up a string into individual characters. In the example that follows, split breaks the protein sequence apart and stores each symbol into the cells of the array @proteinSymbols.

```perl
# split on the empty pattern; returns individual letters

$proteinSequence = "VWFAMILY"; # hydrophobic AAs

my @proteinSymbols = split ( //, $proteinSequence );
```

@proteinSymbols

V	W	F	A	M	I	L	Y

9.8.5 Summary of Using `split`

Delimiter	Pattern	Example use
whitespace	`" "`	`my @items = split( " ", "one two three");` # $items[0] is "one" # $items[1] is "two" # $items[2] is "three"
comma	`/,/`	`@items = split( /,/, "one, two three");` # items[0] is one # items[1] is two three
colon	`/:/`	`@items = split( /:/, "one:two:three, four");` # items[0] is one # items[1] is two # items[2] is three, four
pipe	`/|/`	`@items = split( /|/, "one, two|three");` # items[0] is one, two # items[1] is three
start of line	`/^/`	`@items = split( /^/, $manyLines);` # items[0] is first line in $manyLines # items[1] is second line in $manyLines # etc
nothing	`//`	`@items = split( //, $sequence);` # items[0] is first character in $sequence # items[1] is second character in $sequence # etc

9.9 Using `join`

`join( DELIMITER, LIST )`

The `join` function joins all the individual elements stored within a LIST (or cells of an array) into a single string with each item separated by a specific DELIMITER. Note that unlike the `split` function, `join` does not have the regular expression pattern as the initial argument.

9.9.1 Preparing Tab-Delimited Output with `join`

One example where we like to use `join` involves preparing a list of items to print to an output file where we want each of the individual elements to be separated by a tab (`\t`). As we've seen before, tab- or comma-delimited output is especially useful when you will open the output of your Perl program with another tool, such as the Excel spreadsheet program.

```perl
# assume filehandle $TO_EXCEL is already established

# join multiple values into one tab-delimited string to print
# to Excel

my $line;

$line = join( "\t", $genus, $species, $percent_G, $percent_C );

print $TO_EXCEL "$line \n";
```

9.9.2 Converting an Array of Characters into One String

The `join` function makes it easy to create a string from individual characters stored in an array. Each character stored in the cells of the array `@shuffledNucleotides` will be joined (with nothing, the empty string) into one string.

```perl
# join individual letters into a sequence

$randomSequence = join ('', @shuffledNucleotides);
```

9.10 Sorting Arrays

sort *[USER_SUBROUTINE_NAME] LIST*

The `sort` function arranges the values found in LIST and returns the list in sorted fashion. By default, the `sort` function uses standard string comparisons to determine if one item should appear before another. Thus when the items you want sorted are not strings, additional care must be taken.

9.10.1 Sorting Arrays of Strings

The most straightforward use of `sort` is the sorting of string values within an array.

```
@patronSaints = ( "Kent", "Chargaff", "Wall", "Franklin" );

@patronSaints = sort @patronSaints;
```

9.10.2 Sorting Arrays of Numbers

Sorting items other than simple strings requires extra care, even if the items are as simple as integers. Assume you have an array that holds one of the many results of a microarray analysis for thousands of genes. For demonstration purposes, suppose our focus is on the red fluorescence intensity for each of the first five genes on the chip. We would like to sort these intensities.

```
my @redSignals = (1500, 1246, 1099, 1003, 2798);
```

Recall that the relational (or comparison) operators for comparing two strings (eq, le, gt, etc.) are not the same operators to use when comparing numerical values (==, < =, >, etc.). Given that the default behavior of sort is to assume it is sorting strings, the function will by default use the string comparison operators. Thus, if we want to sort an array of numerical values, we have some extra work to do.

When you want to change the default sorting behavior (and that is almost always the case, the only exception is when you are sorting simple strings), you must write a new subroutine to define how to compare two elements to determine if one should come before the other. The name of the subroutine that you write to do this becomes the first argument, *USER_SUBROUTINE_NAME*, in the sort function. The subroutine must return an integer less than, equal to, or greater than zero depending on how you want the elements in the list or array to be ordered. A sample subroutine to order the elements from smallest to largest is shown below. *Note:* This subroutine is named numerically_the_long_way to indicate that a shorter version will be introduced once we understand how it works.

```
# not the shortest solution, but semantically obvious
# This subroutine is used by sort when comparing two
#       numerical items to determine their numerical order.
# Note: the two items are automatically stored in $a and $b,
#       so do not attempt to declare $a and $b

sub numerically_the_long_way
{
    if ( $a < $b )
    {
        return -1;    # means $a should come before $b
}
```

```
        elsif ( $a > $b )
        {
            return 1;      # means $b should come before $a
        }
        else # $a == $b
        {
            return 0;      # otherwise the two items are the same
        }
    }
```

The use of $a and $b in the numerically_the_long_way subroutine requires some faith on your part. At this point, it would be reasonable for you to ask, "But where are $a and $b declared? From where do the values in $a and $b come? Why does returning a minus one (−1) signal that the value in $a should come before the value in $b?" Good questions.

When you call Perl's sort function, it repeatedly selects two items from the list or array to compare and assigns those two values to the globally-defined variables $a and $b. "Globally defined" means the variables $a and $b are defined elsewhere, specifically in a built-in module (see chapter 14) containing the sort function. The variables $a and $b are actually aliases to the real values, and you should never try to change their values. The sort function calls your subroutine repeatedly, each time setting the next two values to compare into $a and $b. For every two values sent to your subroutine, the sort function waits for a return value to indicate the result of the comparison. The use of −1, 1, and 0 for return values are part of a standard definition when comparing two items; you cannot alter this definition when using sort. If $a is less than $b, sort *expects* that your subroutine will return a −1. In short, your subroutine (e.g, numerically_the_long_way) defines when one element should precede another element and returns the appropriate integer.

Perl's comparison relational operators, < = > for numbers and cmp for strings, automatically return −1, 1, and 0 according to the standard definition of comparing two items. Thus we can rewrite the subroutine to be used by sort when sorting numerical items by relying on the < = > operator rather than the if-elsif-else control structure.

```
# This subroutine is used by sort when comparing two
#       numerical items to determine their numerical order.
# Note: the two items are automatically stored in $a and $b,
#       so do not attempt to declare $a and $b

sub numerically_increasing
{
        return ( $a <=> $b );
}
```

Given the subroutine numerically_increasing, you can use this subroutine name as the first argument when using sort. Notice the use of the subroutine name when calling sort.

```
my @redSignals = (1500, 1246, 1099, 1003, 2798);

my @sortedValues = sort numerically_increasing @redSignals;

print "Sorted: @sortedValues \n";
```

Sorted: 1003 1099 1246 1500 2798

You may have already surmised that if you want to sort numbers in decreasing order, you can reverse $a and $b as shown here.

```
sub numerically_decreasing
{
        return ( $b <=> $a );
}
```

It is worth mentioning that it is possible to avoid the subroutine entirely and include the comparisons inline. In fact, as we'll see later in chapter 10, performing the comparison inline without a subroutine is sometimes the easiest way.

```
my @redSignals = (1500, 1246, 1099, 1003, 2798);

# sort "inline" without using a subroutine

my @sortedValues = sort { $a <=> $b } @redSignals;

print "Sorted: @sortedValues \n";
```

9.10.3 Sorting Arrays by Specific Fields

More often than not, the data you would like to sort is not of the form of basic strings or numbers. No worries. Now that you can write your own subroutine to establish your own definition for comparison and sorting order, you can derive useful parts of $a and $b, compare the parts, and return the appropriate value of -1, 1, and 0. For example, suppose you want to sort an array of lineages for many prokaryotic genomes. Assume that each lineage is prefaced by the GenBank accession number and the number and levels of the lineage are delimited by semicolons. The following is an example for one genome.

```
'AE000512;Bacteria;Thermotogae;Thermotogales;Thermotogaceae;
Thermotoga;Thermotoga maritima MSB8'
```

Assume each lineage is stored in one cell of an array. The initial three cells of the array @genomeLineages are shown next.

'AE000512; Bacteria; Thermotogae; . . .'	'AE009950; Archaea; Euryarchaeota; . . .'	'BA000002; Archaea; Crenarchaeota; . . .'
0	1	2

To sort by the initial two levels of taxonomic names, we need to isolate the domain name (the second field) and group name (the third field) using split. By using both fields, we can ensure that all Archaea genomes and all Bacteria genomes are alphabetically sorted by their next-level taxonomic name, respectively.

```
#--------------\
# taxonomically \
#-----------------------------------------------------------
# This routine is called by Perl's sort() function
#
# This rourtine sorts first by domain name if those are identical
# it will sort by the second level (group) names
#
# IN: two semicolon delimited strings of lineage (stored in
#     $a and $b) e.g.: "AE000512; Bacteria; Thermotogae;..."
#
# RETURNS:   -1 if $a comes before $b
#             1 if $b comes before $a
#             0 if $a is the identical lineage of $b
sub taxonomically
{
    # isolate the fields to sort by [0] = Accession [1] = Domain
    # [2] = Group
    my @a_fields = split( /;/, $a );
    my @b_fields = split( /;/, $b );
```

```
     # we are sorting strings so use the string operators (lt, gt, eq)

     # attempt to sort by first level (Domain)
     if ( ($a_fields[1] lt $b_fields[1]) )
     {
          return -1; # means $a should come before $b
     }
     elsif ( ($a_fields[1] gt $b_fields[1]) )
     {
          return 1; # means $b should come before $a
     }
     else # must be eq, that is, ($a_fields[1] eq $b_fields[1])
     {
          # otherwise the two Domain names are the same
          # so we need to compare by the [2] Group field
          # rather than use another if-elseif-else, we'll use cmp
          # which does all that work for us
          return
               ( ($a_fields[2] cmp $b_fields[2]) );
     }
} # end taxonomically
```

Notice to sort all the Archaea and all the Bacteria alphabetically, we must define a comparison on more than one field. In this case, if the Domain fields are the same (see the else case), we need to refer to the next-level Group name to resolve the comparison. For example, when comparing two Archaea lineages such as Archaea; Crenarchaeota and Archaea; Euryarchaeota, $a_fields[1] and $b_fields[1] both hold Archaea, thus $a_fields[2] and $b_fields[2] are needed to resolve the comparison. Recall that Perl's comparison relational operators (<=> for numbers and cmp for strings) automatically return −1, 1, and 0 according to the standard definition of comparing two items. Thus we have used the cmp operator when comparing the Group names, although you may personally find it more readable to write another set of if-elsif-else statements to compare the Groups.

9.11 Arrays and Regex: Percentage of Hydrophobic Amino Acids in a Protein Sequence

```
$protein = "MTTSMQPSKYTGLVADLMPNIRAMKYSGLFMHNFTGGSAFMKKVYSSV";

printSequence( $protein, 30 );

my $percent;
```

```
$percent = percentHydrophobic( $protein );

printf ("Percent of Hydrophobic Amino Acids: %6.1f \n", $percent);



#------------------\
# percentHydrophobic \
#---------------------------------------------------------------
# Compute the percentage of hydrophobic amino acids
# in the sequence. Note: [VWFAMILY] = hydrophic amino acids
#
# IN: a protein sequence
#
# RETURNS: %hypdrophic amino acids in entire sequence
sub percentHydrophobic
{
    my ($protein) = @_; # a terse shift
    my $AA_length = length( $protein );
    if ($AA_length <= 0)
    {
        print "Warning: amino acid sequence is empty! \n";
        return 0;
    }
    my @allMatches; # array to hold each of the matches
    my $howMany;    # total number of items in the array

    @allMatches = ( $protein =~ m/[VWFAMILY]/ig );
    $howMany = scalar( @allMatches );
    return ( ($howMany / $AA_length) * 100 );
}
#---------------------------------------------------------------
```

9.12 Parallel Arrays

One common convention of handling multiple types of information for many
instances is to use parallel arrays. When arrays are lined up in parallel, cells with the
same index value represent information that is related. For example, suppose you
have access to the results of a microarray analysis involving thousands of genes and
from the results you have stored for each gene the respective red and green fluorescence intensities as well as the red:green ratio. If red signal intensities are stored in
one array, the green signal intensities in a second array, and the red:green ratios in a
third array, a common index across all three arrays provides access to a related set of
information.

```
my @geneIDs        = ( "NC_01",  "NC_02",  "NC_03",  "NC_04" );
my @redSignals     = ( 1500,     1246,     1099,     1003    );
my @greenSignals   = ( 1469,     1554,     1188,     2789    );
my @redGreenRatios = ( 1.02,     0.80,     0.93,     0.36    );
```

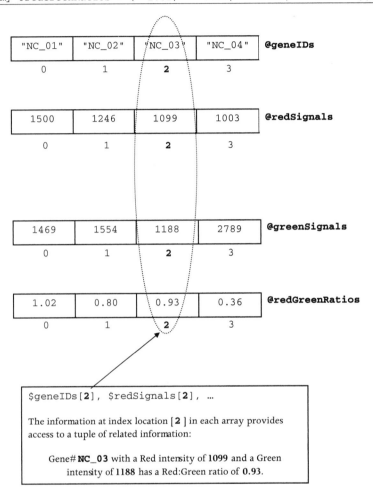

Given a situation where your data are stored in parallel arrays (and you may have many more arrays than just four), one utility you will need is a method to locate the index of a particular tuple of information. For example, if you have an array of thousands of unique gene IDs, how can you find the tuple of information for a particular gene ID number? Put differently, what is the index where a particular gene ID number is located? If you know the index where the particular ID is found, then because the arrays are lined up in parallel, you'll also know the index of all the other pieces of information. If you search for the gene ID NC_03 and you find it at index [2] in the GeneIDs array, then you know that index [2] holds the relevant information in the other arrays. The following subroutine, findGeneID, allows you to send two arguments: (1) an ID you would like to look for (sometimes called the key) and (2) the array of GeneIDs to search. If the subroutine finds the key, it returns the index number where that key was found; otherwise, if the key is not in the array of GeneIDs, the subroutine returns a −1, which means not found. Searching for a key within an array with the following algorithm is known as a linear search.

```perl
my $lookFor = "NC_03";

# search for NC_03 in the array @geneIDs
my $whereFound = findGeneID( $lookFor, @geneIDs );

if ( $whereFound >= 0)
{
    # found it! info is at [$whereFound] for each array
    print "GENE ID:        $geneIDs [$whereFound] \n";
    print "Red Signal:     $redSignals[$whereFound] \n";
    print "Green Signal:   $greenSignals[$whereFound] \n";
    print "Red:Green ratio: $redGreenRatios[$whereFound] \n";
}
else
{
    print "The ID $lookFor is not in the array of Gene IDs \n";
}
```

```perl
#----------\
# findGeneID \
#-------------------------------------------------------------
# SUMMARY: Subroutine to perform a linear search of the array
# @IDs for the value stored $findThisID.
#
# IN: $findThisID is the value to find
#     @IDs is the array of values to search
#
# RETURNS: the index where $findThisID is found within
# the array @IDs or -1 if not found
sub findGeneID
{
    my ($findThisID, @IDs) = @_; # shift twice
    my $lastIndex = $#IDs; # save the last index number
    my $i = 0;
    my $found = 0; # FALSE, we haven't found it yet
```

```
    while ( (!$found) and ($i <= $lastIndex) )
    {
        if ( ($IDs[$i] eq $findThisID) )
        {
                $found = 1; # TRUE, found it
        }
        else
        {
                $i++; #move to next cell in array
        }
    } # end while not found yet

    if ($found)
    {
        return $i;
    }
    else
    {
        return -1;
    }
}
```

Going Back for More

1. Create a new Perl program (e.g., `globTest.pl`). In the same directory as your Perl program, create a new directory called `Tests`. Create a number of new files. The files do not need to contain anything at this point. Change the file extensions on some of the files in the Tests directory, e.g., change `one.txt` to `one.fna`. Have your Perl program glob only those files that have a `.fna` file extension into an array. Print the size of the array and the file names of all files with a `.fna` extension.

2. Create FASTA format sequence files (e.g., perhaps of coding sequence found at NCBI) and move these files into your Tests directory. Alter your Perl program to open each of the `.fna` files that you glob and *print the header line* of each sequence file. (You may need to revisit chapter 8 to review how to open and read from a file.)

3. Modify no. 2 above to split the header line for each file and print the NCBI identification and reference numbers for each `.fna` file.

   `>gi|`**49175990**`|ref|`**NC_000913.2**`|` `Escherichia coli K12, genome`

4. Create an array of strings and sort it in alphabetical and then reverse alphabetical order. Do the same with an array of integers.

5. Create an array of colon (:)–delimited strings and sort the array by various fields. For example, if each element in the array is of the general form:

Genus:Species:Size

sort the array (i) by genus, then (ii) by species, and finally (iii) by size.

6. Review the subroutine `percentHydrophobic` in section 9.11. Write a subroutine to accept an amino acid sequence and return the percentage of amino acids that are not hydrophic amino acids. Assume the set `[VWFAMILY]` defines the hydrophobic amino acids.

....

10   Hash Tables

In which data are referenced by key words to collect and manage large collections of sequences and their related information in the versatile, table-like data structure of a hash.

From a good index it is possible to find out rapidly what is covered in a book [genome], which subjects are covered in the greatest depth, and to get a "feel" for the shape and style. Experience of unreliable, or tendentious, or badly planned indexes causes us to see slowly just what skill and intelligence go into a good one. Later still, one comes to admire the index as a work of art.

—Byatt, 2001, p.11

Hash tables offer a programmer the ability to intuitively and efficiently store and access related pieces of information. In contrast to an array, which is a list of cells each accessed by an integer, a hash table references its cells by words. For sequence analysis, this is a considerable advantage. To print the count of the number of times a motif occurs in a sequence, a programmer wants to look up the motif by name (e.g., CAAT) rather than a number (motif #39). To convert a codon to its associated amino acid name, a programmer wants a mechanism to quickly map a codon (e.g., GGU) to its name (glycine). Hash tables are good starting places for annotation projects in which data are accumulating about a particular sequence: its locations, its upstream genes, its putative functions, and its relationships with other sequences.

10.1   Motivation for Using Hash Tables

A hash table is an unordered set of key-value pairs where key words (strings) reference their associated values (scalars). The hash table (or hash) data structure is similar to the array data structure in that they both hold a collection of values. A main difference between hashes and arrays is that each cell in a hash table is referenced or indexed by a key (string), whereas cells in an array are indexed by integers. The diagram below shows the Perl declaration of a hash to help keep track of the number of occurrences of each of the possible 256 motifs of length four (4-mers). The diagram assumes the hash has already been populated with (key, value) pairs, in this case (motif, count) pairs.

```
my %motifCount;    # hash table (key, value pairs of the
                   # number of occurrences of each 4-mer
```

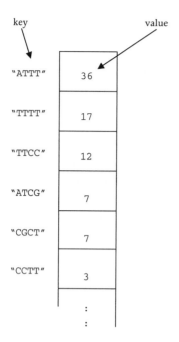

key   value

"ATTT"   36

"TTTT"   17

"TTCC"   12

"ATCG"   7

"CGCT"   7

"CCTT"   3

Alternatively, we could use an array to hold these motif counts, yet an array is not able to facilitate the mapping from motif (string) to count (integer). Recall that when using arrays, a particular cell in the array must be referenced by an integer:

```
my @arrayOfMotifCounts;
$arrayOfMotifCounts[ 0 ]      the initial cell in the array indexed by the
                              integer zero
$arrayOfMotifCounts[ $i ]     the ith cell in the array where $i holds some
                              integer
```

In an array, it would not be clear which motif-count pair would be represented by a particular cell, for example, which motif's count is stored at index zero, $arrayOfMotifCounts[0]? With hash tables, references to particular cells can be handled with the more intuitive key (e.g., ATTT).

A final but significant point is that a hash table provides efficient and fast access to a data item of interest. As we've seen in the previous discussion of parallel arrays (see section 9.12), we could store our motifs in one array and the respective motif counts in another, parallel array. However, arrays are indexed by integers (not strings or motifs in this case), and thus there is no direct way to access the count for a particular motif. If you wanted to update the count for TATA, your program would first have to search through the array of motifs until it found TATA and then, having found the correct cell, the parallel array of counts could be incremented. Functionally, the parallel array with search gets the same task done as a hash table, but the primary value of the hash is speed of access of a particular item.

Box 10.1 The Convoluted Etymology and Use of *Octothorpe* and Its
              Many Synonyms

From the very first Perl program, the symbol # has been used to designate
the part of the program that is not read as code but contains comments in the
programmer's own words.

The symbol # is sometimes called a hash, especially by programmers. And
hash may be shortened to *sh*, as in the pronunciation of "sh bang" for #!, first
symbols of a Perl program. (Note that "bang" charmingly refers to the excla-
mation point.)

Of course, because # (hash) was selected as the comment indicator in Perl,
it could not also be employed as a symbol in building hash tables. For that, %
comes into play. (The symbol % also appears as a modulus operator among the
arithmetic functions.) The word *hash* for # is just one of a long list of potential
names. These have been researched by Michael Quinion for "World Wide
Words," findable by an Internet search.

The # is variously called

- "Pound" because of its association with weight in pounds.
- "Number," perhaps related to its use for weight.
- "Hash," which may be derived from a mispronunciation (from the
  1970s) of *hatch* as in cross-hatching or cutting as with a hatchet, and
  this, too, is the derivation of hash table.
- "Tic Tac Toe" for its pattern.
- "Crunch," perhaps also pattern-like.
- "Square," which was suggested as an international standard because it
  is easily translated into other languages.

Octothorpe seems to have originated as a joke term from Bell Labs in the
1960s, although it did not appear in print until the mid-1970s. They were
adding two new symbols to the telephone dial, the asterisk *, which they
renamed star, and the #, which they called octothorpe. *Octo* is obviously for
the eight points of the figure. The suffix *thorpe* has a few explanations, the
most likely of which may be this, recounted by a retiring Bell Labs scientist:
Apparently at the same time as seeking a fresh new name for the old pound
sign, there was a movement afoot to support the return of Jim Thorpe's
Olympic medals (after a disqualification) and a sympathetic Bell Labs employee
commemorated the name in the newly coined jargon.

10.2 A Simple Hash of Stop Codons

Hashes provide an elegant way to translate three-letter RNA codes to amino acid
symbols or names. As a small introductory example, consider how we might use a
hash table to help translate the stop codons (UAA, UAG, UGA) to their full stop codon
names (Ochre, Amber, Opal, respectively). The three-letter RNA codes (strings)
will serve as the key and the stop codon names as the values. Entries in hashes are

represented as (key, value) pairs, so ("UAA", "Ochre") could serve as one (key, value) pair in the hash table relating RNA codons with stop codon names. In the examples that follow, we'll present a richer example of a hash table that relates any codon to its proper amino acid or stop codon name. For now, notice how hash tables allow the use of a string (the key) such as UAA to reference (find) the associated stop codon name.

$stopCodon{"UAA"}   *the cell in the hash keyed by the codon "UAA"*

"Ochre"   *evaluates to (references) the value 'Ochre"*

The following Perl statements declare a hash table (**%stopCodon**) and explicitly load the (key, value) pairs comprised of (RNA codon, stop codon name).

```
my %stopCodon;        # hash table of (stop codon, stop name) pairs

%stopCodon = (        "UAA" => "Ochre",
                      "UAG" => "Amber",
                      "UGA" => "Opal"
              );

my $nextRNA = "UAG"; # assume "UAG" is next codon to process

print "\n\n";
print "$nextRNA translates to $stopCodon{$nextRNA} \n\n";
```

The output is shown below.

```
UAG translates to Amber
```

Box 10.2 Chargaffian Clues . . . or Things That Don't Add Up

....

One of the puzzles in the deciphering of Egyptian hieroglyphics was that the numbers just didn't add up. On the Rosetta stone were 14 lines of hierogyphics corresponding to 18 lines of Greek. The Greek section had about 500 words, but the hieroglyphics had about 1,400 signs or glyphs. The glyphs were not simple pictograms with a one-to-one correspondence with words. Furthermore, each glyph was comprised of signs. In the entire text, only 66 signs were used over and over in various combinations. Yet these appeared not to be used phonetically, like individual letters and sounds. The puzzle was solved with the realization that the glyphs are combinations (often highly redundant), or *both* phonetic *and* pictorial information.

What does this have to do with DNA? Many numbers and facts about DNA sequences do not add up (yet). The first and most notable numerical problem was Chargaff's numbers. Dismantle a DNA double helix and count the component bases. The frequency of As will be equal to the frequency of Ts; the frequency of Cs will be equal to the frequency of Gs, which is a Biology 101 lesson about DNA. At one point, however, the significance was lost on many researchers including Watson, Crick, Wilkins, Franklin, and even Chargaff himself. Rebuilding a crooked and bumpy DNA model, using Chargaff's numbers, was the eureka moment in early 1953 for Watson and Crick. Suddenly the model helix became a smooth cylinder as bases were aligned and linked in proper symmetry and Chargaff's numbers were transformed into the canonical pairings of A:T and C:G.

Other such problems might be considered "Chargaffian clues." These are situations that do not quite make sense or add up and which therefore suggest that data are missing or alternative points of view are still forthcoming. We include on the list the following examples from research on promoter regions.

To summarize all of these, the lack of extensive conservation of promoter and enhancer motifs does not make sense if the goal is to think of them as linear, vectorial two-dimensional information, a paradigm that has worked well for genes.

1. There are many well-studied and seemingly well-positioned and essential short (10 bp or fewer) motifs in gene promoters. Much lab research in gene regulation has been focused on those. Yet binding proteins often cover much more than 10 bp, and nonspecific, fuzzy binding across longer sequences may be significant.

2. Many lab assays for gene regulation are about turning on or off a particular gene by inserting or deleting a motif in the promoter. However the promoter information may be highly modular and interchangeable. It may be possible to form syntactically correct and functional "sentences" in the promoter that are nonetheless nonsense. We have called this the Mad Libs paradox, after the children's game in which humorous, syntactically and grammatically correct sentences may be formed by inserting various words. The result might be nonsense, such as "The tree ate my uncle."

3. The search for a minimal functional size or consensus sequence for a particular promoter motif may yield results but may not reflect the real use of a particular motif. We have called this the classified ad or Vanna White paradox. Note that "3 bdrm apt" is readable and might represent a minimal functional size of readability in a classified ad. Furthermore, if Vanna White reveals EL_PH_NT, perhaps we see the minimal functional size of the full word *elephant*. However by focusing on minimal functional size in a genome (or in any language) the subtle nuances of

continued

Box 10.2 Continued

grammar and spelling may be overlooked. If the goal is to decipher promoter information, minimal functional size may be misleading.

4. Many mutations in promoter regions have no discernible effect, within the limits of lab assays. Therefore, many promoter sequences seem unconserved and irrelevant. However, the promoter regions are much less dependent on linear, vectorial code. Promoters may be error-tolerant due to redundancies that compensate for missing or misspelled information. Furthermore the spacing of motifs should be considered as informative as the motifs themselves.

5. Enhancers are often considered in a separate category from promoters, being hundreds to thousands of bp away from the gene. Furthermore, for convenience, some promoter studies deliberately focus only on sequences within a few hundred bases of a gene. Enhancers seem to be quite modular and context-sensitive. Including them into more promoter studies might facilitate deciphering regulatory information.

6. Probably all proteins have a minimum, low-affinity binding to DNA. Mutational studies of promoter motifs might shift affinity along a continuum from strong to random. However it should not be assumed that strong binding is representative of actual use in promoting genes. There may be situations in which a lack of or low-level binding forms some essential bit of information, but it would be extremely difficult to detect these above background noise of random binding.

7. Instructions for gene regulation may be scattered throughout the genome. The linearity of chromosomes is an illusion of databases and models. Actual DNA is stuffed into a nucleus or cell such that the information is being used three-and four-dimensionally. Attempting to "read" promoters linearly may never lead to full deciphering of the information.

8. Motifs in promoters and enhancers are often collected, reported, and stored as linear and vectorial (like genes). This is a practical matter for designing databases of sequences, however, it may not reflect the true functioning of motifs.

9. Evolution is quite different from design. It is messy, illogical, and good enough (even if just marginally) for the particular context of a generation in an environment. Therefore, looking carefully for logical patterns may lead to some misunderstandings about promoters.

So what can we do about all of this? Take it as a challenge. We think of the upstreams (and downstreams) of genes as the whitewater of the genome. Like the exploration of whitewater on a raft, we do not expect a linear, vectorial ride on the river. The design of databases to handle intergenic (whitewater) sequences is a major unsolved problem requiring many more dimensions than required for gene data. (Introns, the noncoding sequences within genes, also are not recorded in proper dimensionality, reflecting their complexity.)

Anyone tackling the project might look beyond simple English lexicons as a model. For example, dictionaries of Chinese characters are more complex, sorting by types of brush strokes, numbers of brush strokes, and use of the four corners of the ideogram, as well as phonetics and meaning. Although this probably does not approach the complexity of gene regulation, it might be a better model by which to jump-start the process of designing a database in the mind of an intrepid and adventurous programmer.

Another creative model might be the structure of poetry or music (or the two together) that depend on so much more than linearity. Repetitions, rhythms, rhymes, stress, and accents play into poetry as well as, of course, context and meaning. Describing the complexity of music here would be too much of a digression except to say that its complexity may approach what is going on with gene regulation.

10.3 Hash Tables in Memory

With hash tables, you are not guaranteed that the order in which (key, value) pairs get inserted will be identical to the order stored in the hash. The order in which (key, value) pairs are loaded into a hash table depends on the internal (hidden from us) hashing algorithm used by the Perl implementation. For the time being, rest assured that the (key, value) pairs do indeed get loaded and can be retrieved in a very efficient (fast) fashion.

10.4 Working with a Hash of Amino Acids

The following statements show a complete (RNA, amino acid name) hash table initialization.

```
my %amino_acid =
  (
    UUU => "Phenylalanine",      AUU => "Isoleucine",
    UUC => "Phenylalanine",      AUC => "Isoleucine",
    UUA => "Leucine",            AUA => "Isoleucine",
    UUG => "Leucine",            AUG => "Methionine, START",
    UCU => "Serine",             ACU => "Threonine",
    UCC => "Serine",             ACC => "Threonine",
    UCA => "Serine",             ACA => "Threonine",
    UCG => "Serine",             ACG => "Threonine",
    UAU => "Tyrosine",           AAU => "Asparagine",
    UAC => "Tyrosine",           AAC => "Asparagine",
    UAA => "Ochre (STOP)",       AAA => "Lysine",
    UAG => "Amber (STOP)",       AAG => "Lysine",
    UGU => "Cysteine",           AGU => "Serine",
```

```
        UGC => "Cysteine",              AGC => "Serine",
        UGA => "Opal (STOP)",           AGA => "Arginine",
        UGG => "Tryptophan",            AGG => "Arginine",
        CUU => "Leucine",               GUU => "Valine",
        CUC => "Leucine",               GUC => "Valine",
        CUA => "Leucine",               GUA => "Valine",
        CUG => "Leucine",               GUG => "Valine",
        CCU => "Proline",               GCU => "Alanine",
        CCC => "Proline",               GCC => "Alanine",
        CCA => "Proline",               GCA => "Alanine",
        CCG => "Proline",               GCG => "Alanine",
        CAU => "Histidine",             GAU => "Aspartic acid",
        CAC => "Histidine",             GAC => "Aspartic acid",
        CAA => "Glutamine",             GAA => "Glutamic acid",
        CAG => "Glutamine",             GAG => "Glutamic acid",
        CGU => "Arginine",              GGU => "Glycine",
        CGC => "Arginine",              GGC => "Glycine",
        CGA => "Arginine",              GGA => "Glycine",
        CGG => "Arginine",              GGG => "Glycine"
    );
```

We can use the %amino_acid hash to highlight a number of issues pertaining to the use of hashes.

10.4.1 Syntactic Features Unique to Hash Tables

1. References to the *entire* hash table are prefaced with the percent symbol (%).
2. The operator (=>), sometimes called the "fat comma" operator, is useful when initializing hash table (key, value) pairs. If you know the (key, value) pairs ahead of time, we recommend this syntax for initializing pairs in the hash: key => value. Notice that even though the keys are strings, you do not have to quote the keys. The fat comma operator assumes your keys are to be quoted.
3. Like arrays, when referencing a single (scalar) element in the hash, we use the dollar ($) syntax prior to the name, **$amino_acid**, but the key is enclosed within open and closing braces {key}. Thus, to reference ("find", "look-up") the amino acid name for the start codon AUG, we need not worry about where the cell is located in the hash table, rather we can directly use AUG as the index:

$amino_acid{"AUG"}

which results in the value: "Methionine, START".

....

Good Practices

Notice that we use names in the singular for hashes (e.g., %amino_acid, %stopCodon). Because most references to a hash table in your programs refer to a single (key, value) pair, this naming convention can help facilitate the reading of your programs. Alternatively, because arrays are often referenced collectively in your program, we use plural names for them (e.g., @bacteriaFilenames, @patronSaints).

....

10.4.2 Accessing All the Values in the Hash

Perl provides two functions to iterate over the entire set of keys in the hash, one at a time. The functions are each and keys. The each function steps through the (key, value) pairs one at a time and returns either a two-element list consisting of the next (key, value) pair

```
($key, $value) = each %hash_table; # next (key,value) pair
```

or just the next key.

```
$key = each %hash_table; # next key
```

A while loop is shown below using each to capture each (key, value) pair and print the list of 64 codons, one per line. When the entire hash has been read, the each function returns an empty list that evaluates to false in the while loop test.

```
# print all (key, value) pairs in a hash

my $codon;      # next key
my $AAname;     # next value
my $n = 1;      # count amino acids

print "Internal (key,value) pairs \n";
while ( ($codon, $AAname) = each %amino_acid )
{
   print "($n) $codon - $AAname \n";

   $n++;   # add one to count
}
```

A partial output of printing (key, value) pairs is shown next. Again notice that the internal order of (key, value) pairs is not the same as the order of initialization.

```
Internal (key,value) pairs
(1) GCC - Alanine
(2) UAA - Ochre (STOP)
(3) CGU - Arginine
(4) CGA - Arginine
(5) CUU - Leucine
(6) GAU - Aspartic acid
(7) AAC - Asparagine
(8) AGC - Serine
(9) CUC - Leucine
(10) UCA - Serine
(11) UAG - Amber (STOP)
(12) UUC - Phenylalanine
        :
        :
```

Alternatively, the keys function returns a list of just the keys in the hash. To print the (key, value) pairs in alphabetical order of key, we can sort the list of keys once they are returned by the keys function as shown below. Notice in this example that the keys function returns a list of keys, thus the hash table itself is not sorted, rather a separate and independent list of keys returned from the keys function is sorted. A full reference into the hash using the next key is needed to retrieve the associated amino acid name value (e.g., $amino_acid{$codon}). Because a single access into the hash is theoretically as fast as any other access, asking for (key, value) pairs after the keys have been sorted is as efficient as asking for (key, value) pairs with unordered keys.

```
# print all (key, value) pairs by alphabetically sorted keys

$n = 1;
foreach $codon ( sort keys %amino_acid )
{
    print "($n) $codon - $amino_acid{$codon} \n";
    $n++;
}
```

A partial output of printing (key, value) pairs with alphabetically sorted keys is shown here.

```
Sorted KEYS ...
(1) AAA - Lysine
(2) AAC - Asparagine
(3) AAG - Lysine
(4) AAU - Asparagine
(5) ACA - Threonine
(6) ACC - Threonine
(7) ACG - Threonine
(8) ACU - Threonine
(9) AGA - Arginine
(10) AGC - Serine |
(11) AGG - Arginine
(12) AGU - Serine
        :
        :
```

You may also want to save the sorted key values in an array for later use.

```
# save alphabetically sorted keys into an array

my @sortedKeys;

@sortedKeys = sort ( keys %amino_acid );
        :
        :
```

10.4.3 Sorting a Hash by Values

As shown in section 10.4.2, the keys function in conjunction with the sort function can create an alphabetically sorted list of keys. However, as described in section 9.10, you must write your own comparison statements to sort by a more complicated definition, including the case where you want to sort by the values in the hash. The following algorithm provides an outline of the steps needed to sort a hash by its values.

1. Form a list of all the keys in your hash (e.g., keys %hash).
2. In the block where the sorting of two elements is performed, compare two hash values based on the keys using cmp if comparing strings and <=> if comparing integers. Recall that sorting in increasing order requires a definition of "$hash{$a} comes before $hash{$b}" (e.g., $hash{$a} cmp $hash{$b} if comparing strings), while sorting in decreasing order requires a definition of $hash{$b} cmp $hash{$a}.
3. Store the hash keys in an array; the initial element in the array of keys will hold the key that corresponds to the value in the hash that should come first based on the sorting.
4. Traverse the array of keys to access the hash values in sorted order.

The following program can be used to sort a hash by its values, for example, sorting by the names of the amino acids. Again, note that the criterion for sorting is based on the comparison of the values, but we will store the keys that are associated with those sorted values.

```
# (1) using all the hash keys
# (2) sort by hash VALUEs (use cmp since values are strings)
# (3) store keys in an array

my @keys_sortedByValue;
```

```
@keys_sortedByValue =                                    # (3)
  sort { $amino_acid{$a} cmp $amino_acid{$b} }           # (2)
  keys %amino_acid;                                       # (1)
print "Hash sorted by VALUE . . . \n";
$n = 1;
foreach $codon ( @keys_sortedByValue )
{
    print "($n) $codon - $amino_acid{$codon} \n";
    $n++;
}
```

A partial output of the (key, value) pairs after having been sorted by the values in the hash is shown below. Note that it is the values that are sorted: "Alanine" < "Amber (STOP)" < "Arginine" and so on.

```
Hash sorted by VALUE ...
(1) GCC - Alanine
(2) GCU - Alanine
(3) GCA - Alanine
(4) GCG - Alanine
(5) UAG - Amber (STOP)
(6) CGU - Arginine
(7) CGA - Arginine
(8) AGA - Arginine
(9) CGC - Arginine
(10) AGG - Arginine
(11) CGG - Arginine
(12) AAC - Asparagine
(13) AAU - Asparagine
(14) GAU - Aspartic acid
(15) GAC - Aspartic acid
(16) UGU - Cysteine
(17) UGC - Cysteine
(18) GAG - Glutamic acid
          :
          :
```

10.4.4 Determining the Number of (Key, Value) Pairs in Your Hash

Assigning the value returned from the keys function to a scalar will indicate the number of (key, value) pairs in your hash.

```
my $hashSize;

$hashSize = scalar( keys %amino_acid );

print "The hash has $hashSize (key,value) pairs \n";
```

The output indicating the size of the amino acid hash is shown next. With four RNA nucleotide choices for each of the three slots, the size is $4^3 = 64$.

```
The hash has 64 (key,value) pairs
```

10.4.5 Testing If a Key Already Exists in a Hash

When you want to see whether a particular key already exits in a hash table, the exists function can be used with an if-else conditional statement. The exists function returns true if the specified hash key or array index exists in the data structure. The exists function would return true in the following example because "Rice" is a valid key already in the hash.

```perl
my %geneCount =
    (
            "Tuberculosis microbe"  => 4000,
            "Yeast"                 => 6000,
            "Fruit Fly"             => 13000,
            "Mustard Weed"          => 26000,
            "Human"                 => 30000,
            "Rice"                  => 58000,
    );
if ( exists( $geneCount{"Rice"} ) )
{
    print "Yes, Rice is in the hash table already. \n";
}
else
{
    print "No, Rice is not yet in the hash table. \n";
}
```

10.4.6 Adding a New (Key, Value) Pair to a Hash

Adding new (key, value) pairs to your hash table is easy.

```perl
# add a gene count for Caeno, the nematode worm, to the
# hash

$geneCount{"Worm"} = 18000;
```

10.4.7 Deleting a (Key, Value) Pair from a Hash

```
# remove a (key,value) pair

delete ( $geneCount{"Yeast"} );
```

Box 10.3 Complexity: $2X$ versus. X^2'

A major problem in DNA sequence analysis or any other topic in bioinformatics is complexity. Even the shortest list of elements can escalate into numerous potential combinations and connections. Microbiologist Jared Leadbetter, speaking at a Gordon conference in 2005 about the challenges of sorting out the interactions of a community of bacteria, said, "It is the difference between $2X$ and X^2."

In Perl, we would say:

```
my $X; # X is the number of items we are considering
print "Complexity is not about $X * 2 \n"; # Complexity
                                             isn't 2X
print "Complexity is about $X ** 2 \n"; # Complexity is X²
```

10.5 Counting the Frequency of Unique Motifs

Suppose you would like to count the number of times each unique 4-mer motif appears in a genome and report the top 10 most frequently occurring motifs. (The English language equivalent is to wonder about author Lewis Carroll's most frequently used words in the book *Alice in Wonderland*. *Hint*: The word *the* must be right up there near the top, right?) Let's consider why a hash table is indeed the data structure that we want for counting the number of times each motif appears in a sequence of DNA.

To count each motif, we need an algorithm to work through a genome (or some DNA sequence) one 4-mer at a time. Consider how you would determine the top 10 motifs if you had to do this by hand? As you "read" through the DNA sequence, you might set up a tally sheet like shown here:

motif	how many
AAAA	‖ ‖ ‖
AAAT	⧺⧺ ‖ ‖
AAAG	‖ ‖
⋮	
⋮	

Each time you read a new 4-mer (e.g., AAAG) you'll scan down your tally sheet for this motif and if you find it, you'll count it by adding a new tally mark. If you don't find that motif, this must be the first occurrence of that 4-mer, so you'll add it to the bottom of your tally sheet with a count of one. As it turns out, this is almost exactly what Perl does when you solve this problem with a hash table.

Because we want each unique motif to be *associated* with a single frequency count and we want to reference (key, find, look-up) the count for a specific word by keying on the actual motif name (string), a hash table is the data structure for this job!

The following Perl program breaks a DNA sequence into unique 4-mers, uses each 4-mer as the key into the hash table, and adds one to the count for that 4-mer.

```
# a hash table of frequency counts for each DNA motif

my %motifCounts;

my $LMER = 4;                # let's count 4-mers, L = 4

my $DNA = "GAAAAACAAAAC";    # simulate some DNA sequence
my $DNAlen = length($DNA);

my $startBP = 0;      # take L-mer substrings starting here
my $end = $DNAlen-($LMER-1);      # upto here

while ($startBP < $end)
{
        # snag $LMER nucleotides from $DNA at $startBP
        my $nextMotif = substr($DNA, $startBP, $LMER);

        # add one (++) to this next motif's count
        $motifCounts{$nextMotif}++;

        $startBP++; # move to next motif
} # while more motifs
```

Starting with the DNA sequence GAAAAACAAAAC, the (key, values) in the hash table might possibly look like this in memory.

"GAAA"	1
"CAAA"	1
"AAAC"	2
"AAAA"	3
"AACA"	1
"ACAA"	1

An additional discussion is in order regarding the line of Perl that increments the count for the next motif that was found.

```
# add one (++) to this next motif's count
$motifCounts{$nextMotif}++;
```

A longer version of this statement is perhaps easier to read.

```
$motifCounts{$nextMotif} = $motifCounts{$nextMotif} + 1;
```

If you were to read this statement aloud you might say, "Starting on the right-hand side of the assignment operator (=), find the location in the hash for the next motif and having retrieved its associated count, add one to that count. Then store the newly incremented count into the hash cell associated with this motif." What happens when the hash encounters a key (motif) seen for the first time? Referencing a nonexistent hash value evaluates to zero, thus adding one to a nonexistent hash value will create it with an initial value of one.

Box 10.4 The Three Bears versus The Gettysburg Address

....

The comparison of sequences is a foundation of genomics and proteomics. Many fine software packages have been developed to facilitate sequence comparisons and for building phylogenetic trees. The results and interpretations are the basis for many contentious, unresolved arguments about significance and assumptions. Chances are, you know of at least one controversy over phylogenetics in your own area of expertise. A simple classroom exercise to demonstrate some of the tree-building problems to students is to hand out several strips of paper on which are printed short sequences that differ slightly by either insertions or deletions or base changes. Have each student try to lay out the strips of paper in a family tree pattern, and immediately the controversy of where to root the tree is revealed. Most likely, there will be several versions of trees and different opinions on what to do about insertions or deletions.

A more advanced exercise is the "Tree-Thinking Challenge" (Baum, Smith, and Donovan, 2005). Do an Internet search to find the lengthier online version and try the test on yourself, colleagues, and students. It can be a humbling experience to realize how easily an interpretation may be skewed simply by the artistic or design decisions concerning the format and structure of a tree.

Our book does not venture into the tricky area of advising and instructing on building trees. Many other books focus entirely on that matter. Rather, in the spirit of the way we have been handling most topics of this sort, we suggest some ways to take creative or unconventional approaches. When you design and implement your own software, you have the luxury of being able to work a little outside of the box.

One of our favorite creative approaches we nicknamed "The Three Bears versus The Gettysburg Address." It came about from our attempts to understand intergenic sequences, which do not lend themselves well to the linear, vectorial analyses used for genes. If you were to build a hash of all of the words in the fairy tale "The Three Bears" and all of the words in Abraham Lincoln's Gettysburg Address, you would have two rather different and informative lists about the nature of the English language. The roughly comparable exercise for DNA would be to choose a range of lengths and collect all possible motifs of those lengths in a hash. In the case of the fairy tale versus political speech analysis, you might find that common words like *the* were present in both, while other words or phrases like "Four score" or "Once upon" were rare and unique to each document, part of a signature of authorship. We hypothesized that such an authorship approach might be taken with promoter regions to decipher what words or phrases or combinations might be distinctive and perhaps important and informative. A phylogenetic tree in this context is more of a tree representing the body of work of a particular "author." For example, a set of all of Lincoln's speeches are likely to have more in common within the group than with a set of fairy tales. Software could be designed to analyze and detect the authorship of a novel sequence and to place it into the appropriate category. The work of another author (or a promoter region known to be different) or a randomly generated text would not fall into either set. Taking the linguistic analogy further, we enjoy appropriating some of the vocabulary to talk about motifs. A few examples include:

- "hapax legomenon," a word that appears only once in a text.
- "lemmatize," sorting words as they occur in a text, grouping those that are variant forms of the same word.
- "calque," a word or expression formed from the translation of another word, for example, the French *gratte-ciel* is a calque of *skyscraper*. (Do some horizontal transfer events result in a new use and interpretation of the "foreign" sequence?)
- "false friend," sounds like the word you know but is of completely different derivation (with the constraints of what is chemically possible with just four bases, there must be lots of these in the analysis of short sequences).
- "shibboleth," a word that is easily used and pronounced only by native speakers (sequences that reveal that viruses, for example, are foreign to a cell).

10.6 Putting It All Together: Oligonucleotide Counts for Multiple Genomes

Section 10.5 used a hash table to count the number of words in a single DNA sequence. Of course, rather than a simple DNA sequence, you could download a number of genomes or coding regions from NCBI and count the motifs in each genome using a hash. Because each genome will use many and sometimes all the same motifs, we can use the same hash table for all the genomes with some extra work. Rather than immediately presenting the entire solution all at once, we present some of the steps on the way to a working solution, including documentation and sketches that would appear on the whiteboard as we designed the solution. If you need a review of how to download genomes such as these used in this example, see section 8.6.

> *Summary:* Store the oligonucleotide counts (4-mers) for three prokaryotic genomes, one bacteria, and two archaea: (1) *Escherichia coli* K12, (2) *Methanocaldococcus jannaschii*, and (3) *Pyrococcus furiosus*. Print three separate reports to tab-delimited output files. For each genome, in a separate file, print the percentages of the top 10 most frequently occurring motifs in a file called *genomeName_*Top10.xls, where *genomeName* is dynamically concatenated to be part of the file name. After all genomes are processed, print the top 10 average percentages to STDOUT (the console screen).

> *Input:* Three complete DNA genomes from NCBI, each file in FASTA format.

> *Output:* Three Excel-ready output files. After a complete run, the files will contain the genome name (file name), the size of the genome (bp), and three columns such as shown below.

Ecoli_K12	4639672 (bp)	
Rank	*Motif*	*Frequency (%)*
1	CAGC	0.808
2	GCTG	0.787
3	TTTT	0.7675
4	CGCC	0.7578
5	AAAA	0.7573
6	GCGC	0.7561
7	GGCG	0.7433
8	CCAG	0.7385
9	CTGG	0.7274
10	GCCA	0.6853

Also print the top 10 average frequencies (percentages) over all the genomes to the screen.

> *Data Structure:* A hash of arrays called %motifCounts. A hash table will enable us to quickly reference the correct cell given a particular motif as the key. Because we have multiple genomes and we need composite information over all the genomes (e.g., average motif frequency), each genome will need a record of frequency counts for each motif, thus the data structure we need is a hash of arrays as diagrammed below.

[0] *Escherichia coli* K12 (Bacteria; Gammaproteobacteria*)*

This organism was named for its discoverer, Theodore Escherich, and is one of the premier model organisms used in the study of bacterial genetics, physiology, and biochemistry. This enteric organism is typically present in the lower intestine of humans, where it is the dominant facultative anaerobe present, but it is only one minor constituent of the complete intestinal microflora. (Summary from NCBI)

[1] *Methanocaldococcus jannaschii* (Archaea; Euryarchaeota)

This organism is an obligately anaerobic methane-producing archeon and was the first representative of the archaeal domain to be completely sequenced. It was isolated in 1982 from a deep-sea hydrothermal vent. (Summary from NCBI)

[2] *Pyrococcus furiosus* (Archaea; Euryarchaeota)

This organism is a strictly anaerobic, hyperthermophilic archeon. This organism is highly motile due to a bundle of flagella. This strain was isolated from a shallow marine solfataric (volcanic area that gives off sulfuric gases) region at Vulcano Island, Italy. (Summary from NCBI)

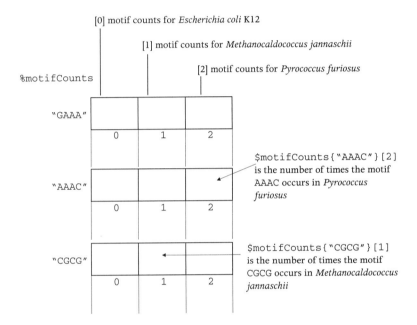

Note the syntax to refer to a particular genome's motif count, For example, because the motif counts for *P. furiosus* are stored in the array of counts at cell [2], if we encountered the motif AAAC, the following line would increment the count for AAAC in *P. furiosus* by one more:

```
$motifCounts{"AAAC"}[2]++;
```

In summary, our data structure, though more complex than what we've seen so far, is an elegant representation of the problem. We need to count motifs so we use a hash table to quickly access the correct location by using the next motif, for example, AAAC. Yet because we are counting the number of times AAAC (and all other 4-mers) that occur in more than one genome, we need an array of counts for AAAC, an array of counts for CGCG, and so on—an array of counts for each motif where the cells in each array refer to the respective genomes of interest. Our data structure is thus a hash of arrays.

```perl
# count motifs in multiple genomes

#!/usr/bin/perl

use strict;
use warnings;
use File::Spec; # module to help locate current working directory

# a hash of arrays of frequency counts for each DNA motif
my %motifCounts;

my $LMER = 4;        # let's count 4-mers

my @genomeName;      # array of genome names

#-----------------------------------------------------------------
# GLOB the input files

# determine the absolute pathname to the current directory
my @path          = File::Spec->splitdir( File::Spec->rel2abs($0) );
my $abs_pathname = File::Spec->catfile( @path[0 .. $#path-1], "" );

my $fullname = $abs_pathname. "genomes/*.fna";

print "fullname path: $fullname \n";

# grab a "glob" of files and put into elements of the array @fileList
my @fileList = glob($fullname);

# = = = = = = = = = = = = = = = = = = = = = = = = = = = = = = = = =
# process the genomes one at a time

my $n = 0; # nth genome in progress

# for each genome
foreach my $nextFile (@fileList)
{
    #-------------------------------------------------------------
    # cut out the filename used for this genome

    $nextFile =~ m/.*\/(.+).fna$/;
    $genomeName[$n] = $1;

    #-------------------------------------------------------------
    # read the DNA from the file into one string

    my $DNA = readInDNA($nextFile);

    #-------------------------------------------------------------
    # count each LMER in the genome
```

```perl
my $DNAlen = length($DNA);
my $startBP = 0; # take L-mer substrings starting here
my $nMotifs = $DNAlen-($LMER-1); # upto here (which is the #
                                 # of motifs)

while ($startBP < $nMotifs)
{
    # snag $LMER nucleotides from $DNA at $startBP
    my $nextMotif = substr($DNA, $startBP, $LMER);

    # add one (1) to this motif's count for nth genome
    $motifCounts{$nextMotif}[$n]++;

    $startBP++;

} # while more motifs
#-------------------------------------------------------------
# create an Excel output file for this genome

my $ExcelFILE;

my $outputFilename = $abs_pathname. "$genomeName[$n]". ".xls";

open ( $ExcelFILE, '>', $outputFilename )
    or die "Cannot open the input file: $outputFilename: $!";

#-------------------------------------------------------------
# sort counts for genome[$n]

my @keys_sortedByValue =
    sort sortByCount (keys %motifCounts);

#-------------------------------------------------------------
# print report for genome[$n]; use tabs (\t) to separate data
print $ExcelFILE "$genomeName[$n]\t$nMotifs (bp)\n";
print $ExcelFILE "Rank\tMotif\tFrequency (\%)\n";

my $lastIndex = $#keys_sortedByValue;
my $topTen = 1;

while ( $topTen <= 10 and $topTen <= $lastIndex)

{

    my $nextMotif = $keys_sortedByValue[$topTen-1];
```

```
      if ( exists( $motifCounts{$nextMotif} ) )
      {
        my $percent=($motifCounts{$nextMotif}[$n]/$nMotifs) * 100;
        printf $ExcelFILE "%2d\t%4s\t%8.4f\n",
                        $topTen, $nextMotif, $percent;
      }
      $topTen++;
    }
  close( $ExcelFILE );
    #------------------------------------------------------------
    # get ready for next genome[$n]
    $n++; # next genome
}      # for each file globbed


# = = = = = = = = = = = = = = = = = = = = = = = = = = = = = = =
Exercise for the Reader:
# Now compute average frequencies per motif using all genomes and
# print top-10 averages to the console.
#
# HINT: use the next column in the hash to store the averages.


# = = = = = = = = = = = = = = = = = = = = = = = = = = = = = = =

#----------\
# readInDNA \
#------------------------------------------------------------
# SUMMARY: Subroutine to open a FASTA formatted file of DNA sequence
# and return as one long string. The function will remove any
# whitespace, ignore lines that begin with a #, and any digits
# that may appear in sequence, e.g., line numbers or base
# pair counts.
#
# IN:        1 argument: name of file holding the DNA
#            (assumed FASTA format that includes a >header line)
#
# RETURNS: DNA as one string in uppercase nucleotide letters
#
sub readInDNA
{
    my ($filename) = @_; # save filename argument in local variable
    my $firstline;       # holds headerline (ignored)
    my $line;            # holds each subsequent line one at a time

    my $FNAFILE;         # create a file handle to the opened file
```

```perl
    open ( $FNAFILE, '<', $filename )
        or die "Cannot open the input file: $filename: $!";
    # read and ignore the header line if found (complain if not)
    if ( !($firstline = <$FNAFILE>) )
    {
      print "Can NOT read header line from the file called: $filename";
      exit();
    }
    my $seq = ""; # continually concatenate each line to end of $seq
    # WHILE not EOF, grab next line and concatenate sequences
    while ( $line = <$FNAFILE> )
    {
        chomp $line;                  # gobble the newline character
        # discard a blank line
        if ( $line =~ m/^\s*$/ )      # if whitespace (\s) start(^)
                                      # to end($),
        { next; }                     # return to while for next line
        # discard comment line (any line starting with a comment(#))
        if ($line =~ m/^\s*#/)
        { next; }
        $line =~ tr/0123456789//d;    # remove all digits
        $line =~ tr/ \t\n\r//d;       # remove any extra whitespace
        $line = uc($line);            # convert everything to upper
        $seq = $seq . $line;          # concatenate new line onto the
                                      # entire text so far

    } # end WHILE not EOF

    close( $FNAFILE );

    return $seq;
} # end subroutine readInDNA
#-----------------------------------------------------------------



#-----------\
# sortByCount \
#-----------------------------------------------------------------
# SUMMARY: Subroutine to sort the nth column of counts in the
#          hash of arrays.
#
# IN:  no arguments, but Note that the hash (%motifCounts) and the
#      current column (nth genome) are used GLOBALLY here.
#
# RETURNS:   <=> values for sorting in decreasing order of the
#            values in the nth column of the hash of arrays
```

Perl for Exploring DNA

```
# $motifCounts{$b}[$n]< $motifCounts{$a}[$n]
# of if these two counts are equal we'll sort by the motif name
# || $a cmp $b (alphabetically for a tie)
#
sub sortByCount
{
    # NOTE: we must be careful! Not all genomes will have every
    # motif (thus $a or $b could be undefined which means these
    # entries do not exist in the hash at all OR a certain genome
    # may not have any of a certain motif which means the value is
    # undefined;
    #
    # in either case, we'll set that particular motif count to
    # ZERO (0)
    if ( !exists( $motifCounts{$a} ) or !defined
        ($motifCounts{$a}[$n]) )
    {
        $motifCounts{$a}[$n] = 0;
    }
    if ( !exists( $motifCounts{$b} ) or !defined
        ($motifCounts{$b}[$n]) )
    {
        $motifCounts{$b}[$n] = 0;
    }
    return
    (
        $motifCounts{$b}[$n] <=> $motifCounts{$a}[$n]
                            ||
                        $a cmp $b
    );
}
```

Going Back for More

1. Hash tables represent a new level of complexity, so you should expect to face a slightly steeper learning curve. Our best recommendation is to practice, practice, practice. We suggest that you work your way through the chapter again, pausing to try the snippets of code, making small changes, and observing the results.
2. Think of at least two new examples for why you'd want a hash table as opposed to just an array or set of parallel arrays. *Hint*: Indexing by words is the key.
3. Complete the code in section 10.6 to compute and print the top 10 average nucleotide frequencies for each motif for all genomes.

11   Phrasing Questions by Writing Algorithms

In which guidelines are provided for designing and outlining programs
with two sequence aligning programs, BLAST and BLAT as examples.

The boundary is the best place for acquiring knowledge.
—Paul Tillich (in Levy, 2003)

11.1 The BLAST Algorithm

In 1990, Stephen Altschul and colleagues published their algorithm for sequence
comparisons. That was the origin of BLAST, the Basic Local Alignment Search Tool,
well-known now to most researchers in molecular biology. The acronym has even
entered into common use as a verb as in, "Let's BLAST that sequence and see what
it is." The impetus for the creation of BLAST was a need for a rapid search tool for
large data sets using inexpensive and ubiquitous desktop computers. The starting
point for building any such program is an algorithm, essentially an outline or recipe
of the procedure by which a particular problem will be solved. Most likely, Altschul,
and colleagues spent many hours diagramming on the whiteboard and sketching on
scrap paper. They may have experimented with several alternative approaches to
the problem that did not succeed. Here is what they finally produced, paraphrased,
simplified, and rendered by us into a list.

1. Input a query string.
2. Break up the query string into smaller "words" of length W.
3. Find all possible words of length W in the database.
4. Compare each query word to each word in the database, using a matrix
   in which exact matches for each letter are scored $+5$ and mismatches are
   scored -4 (e.g., the query word ATGC matched up with the word ATGG in the
   database would earn a score of $5 + 5 + 5 - 4 = 11$).
5. Look at each score and save those that are high enough by some criterion.
6. Try to improve each score by extending the length of the query word by one
   base at a time to see if the score improves.
7. Keep extending the word in both directions, as long as the score keeps improving.
8. Stop when the score no longer improves and report the results.

Box 11.1 An Algorithm for Discovery: Five Simple Principles

1. Slow down to explore. Discovery is facilitated by an unhurried attitude. We favor a relaxed yet attentive and prepared state of mind that is free of the checklists, deadlines, and other exigencies of the workday schedule. Resist the temptation to settle for quick closure and instead actively search for deviations, inconsistencies, and peculiarities that don't quite fit. Often hidden among these anomalies are the clues that might challenge prevailing thinking and conventional explanations.

2. Read, but not too much. It is important to master what others have already written. Published works are the forum for scientific discourse and embody the accumulated experience of the research community. But the influence of experts can be powerful and might quash a nascent idea before it can take root. Fledgling ideas need nurturing until their viability can be tested without bias. So think again before abandoning an investigation merely because someone else says it can't be done or is unimportant.

3. Pursue quality for its own sake. Time spent refining methods and design is almost always rewarded. Rigorous attention to such details helps avert the premature rejection or acceptance of hypotheses. Sometimes, in the process of perfecting one's approach, unexpected discoveries can be made. An example of this is the background radiation attributed to the Big Bang, which was identified by researchers while they were pursuing the source of a noisy signal from a radio telescope. Meticulous testing is a key to generating the kind of reliable information that can lead to new breakthroughs.

4. Look at the raw data. There is no substitute for viewing the data firsthand. Of course, there is no question that further processing of data is essential for their management, analysis, and presentation. The problem is that most of us do not really understand how automated packaging tools work. Looking at the raw data provides a check against the automated averaging of unusual, subtle, or contradictory phenomena.

5. Cultivate smart friends. Sharing with a buddy can sharpen critical thinking and spark new insights. Finding the right colleague is in itself a process of discovery and requires some luck. Sheer intelligence is not enough; seek a pal whose attributes are complementary to your own, and you may be rewarded with a new perspective on your work. Being this kind of friend to another is the secret to winning this kind of friendship in return.

Although most of us already know these five precepts in one form or another, we have noticed some difficulty in putting them into practice. Many obligations appear to erode time for discovery. We hope that this essay can serve as an inspiration for reclaiming the process of discovery and making it a part of the daily routine. In 1936, in *Physics and Reality*, Einstein wrote, "The whole of science is nothing more than a refinement of everyday thinking." Practicing this art does not require elaborate instrumentation, generous funding, or prolonged sabbaticals. What it does require is a commitment to exercising one's creative spirit—for curiosity's sake.

Reprinted and slightly abridged from Paydarfer and Schwarts (2001).

11.2 **Pseudocode**

Algorithms are typically represented in pseudocode, not necessarily a step-by-step list. Pseudocode is a set of instructions that resembles (thus *pseudo*) the organization, tempo, flow, and content of a set of directives from a programming language, such as Perl. Some of the elements that almost always appear in pseudocode include the following.

1. A clear statement in English as to what you expect the program to do. This should include your intended output, what you hope your program will deliver. Dream big! You can always trim back your expectations later. Notice the output comes first. After all, that is why you are writing the program. One way to think about the design of algorithms is that you will be working backward from your desired result. Rough analogies include envisioning a finished sweater in a particular pattern and then working backward to write out the instructions, all of the knits and purls you will need to accomplish that pattern.

2. A statement about what input you need and where it is coming from. For BLAST, a user will insert a query sequence, and you need access to a database against which to check the query.

3. On a chalkboard, try laying out the entire program in boxes, with each box representing a different activity. Connect the boxes with arrows so that it is clear what goes in (and from where) and what goes out (and to where).

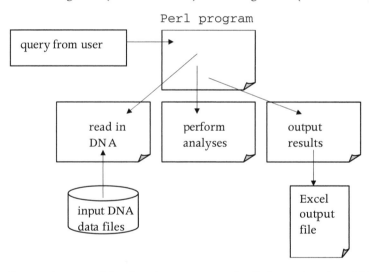

4. Now make a transition from boxes to more detailed pseudocode, which should have familiar Perl-like organizational patterns and conventions. Do this with a liberal use of brackets, including nested ones like { { { } } }, indicating a particular order in which the program will proceed as well as indentations to help you follow the order of the brackets. You've seen and used plenty of

brackets and indentations in the last few chapters. The pseudocode might look something like this:

```
for each word
{
        while (score is improving)
        {
                extend word by one letter
                calculate new score
                compare to previous score
        } end while
} end for
```

5. Variables should be declared and nicely named. However, you are not expected to create an exhaustive list at this stage. More variables will most likely come up later as you solve little problems throughout the program. For example, to write an algorithm for BLAST a partial list of the variables might be

```
$stringToBeSearched
$wordLengthW
$wordLengthWfromDatabase
$TotalScoreForMatches
```

6. Make a well-considered use of preexisting data structures. Make decisions about whether you will use arrays or hashes or whether single variables might suffice to hold your data.
7. Plan a wise use of the many preexisting functions and operations in Perl (or any language you choose). For example, you would be using all sorts of arithmetic operations to calculate and compare scores in the BLAST algorithm. In some cases you would write your own subroutines. It is nice to know ahead of time which functions are look ups, which are built into Perl, which can be lifted from preexisting code and which ones you will have to write de novo. Lifting from preexisting code? Acknowledge in your comments the source of that borrowed code. Some things are needed again and again in various programs.
8. Make careful use of `while` and `for` loops and conditionals, such as `if-elsif-else`, complete with proper brackets, parentheses, and indentations to mimic the flow of the program you intend to write.

When you make the transition from writing the algorithm to writing the actual lines of code:

1. Consider writing the program out of order, beginning with the sections that are either easiest or most intriguing.
2. Consider writing in lots of comments first and then filling in the code. Also set up your choice of decorative lines such as

```
# = = = = = = = = = = = = = = = = = = = = = = = = = = = = =
```

to separate the sections of the program.

Box 11.2 Importance of Beauty I

. . . .

A search for things of beauty is a worthy and time-honored starting premise for scientific exploration. Examples of beauty include elements of novelty, complexity (or simplicity), irregularity (or "perfection"), and surprise. One year we sustained excitement in our genomics research group by writing programs to search genomes for inverted repeats.

ACGTCGAT:ATCGACGT is an inverted repeat because of the internal base pairing. The outside A goes with the outside T, the next C goes with the next to the last G, and so on. The linear sequence can be folded into a hairpin shape and held together by base pairing along the entire length. The rest of the sequence, not in a hairpin loop, is represented here with Ns.

T-A
A-T
G-C
C-G
T-A
G-C
C-G
N N N N N N N A-T N N N N N N N N

At the time we started our research group, we were almost entirely innocent of information concerning the functional relevance of inverted repeats (IRs). One of us (the biologist, of course) was aware of the presence of IRs in introns and in the promoters of some bacterial genes. We still had a lot to learn.

However, what kept us going was the realization that these repeats are beautiful. The more we searched, the broader our definitions of beauty became. It included irregular repeats with mismatches and asymmetrical repeats as well as perfect ones. We soon became intrigued with repeats in general (direct and mirror repeats) and even extended our enthusiasm to English language examples, such as mirror image palindromes. One night we discovered that some DRs are also IRs, such as the lovely GATCGATC; other DRs are also MRs, such as the somewhat less attractive GAGGAG. We then spent considerable time exploring the various properties of repeats, even coining the phrase "versatile repeat" (VR) for the DR/IR and DR/MR combinations.

Beauty is in the eye of the beholder, as we have been told, but there is much to behold in a genome. Somewhere within are sequences that may reveal themselves as beautiful, at least in your own research group. You may want to collect them in a nicely ordered data structure, such as an array or hash. You may want others to appreciate them via friendly search queries and attractive graphics. In cataloging and displaying favorite sequences and making them accessible to others and yourself for further analysis, you are making a much-needed contribution in genomics.

11.3 Writing Algorithms versus Hacking

Beginning programmers often wonder, "Why bother with an algorithm on the whiteboard? Why not just sit down at the computer and start hacking together the program? If it doesn't run immediately, just keep making little tweaks and adjustments until it does." Actually some small programs can be written on the fly, and plenty of programmers write them that way. However a better answer to the question, "Why bother with an algorithm?" is the following.

1. Lots of little experiments (followed by tweaks and adjustments) on a hastily assembled program may end up telling you only about your particular program and your chosen (hastily assembled) data set. You may never discover wonderful alternative, faster, more elegant solutions.

2. Putting together several approaches to a problem and writing them up on the board in algorithmic style gives a group of interested colleagues the opportunity to discuss, compare, contrast, and brainstorm. That is less likely to happen if the group is forced to look over your shoulder as you hammer away on your keyboard.

3. Many worthwhile programs are long, often hundreds to thousands of lines long. It will take days or weeks to produce them in workable form. It is much quicker to write an algorithm and hand trace the steps, looking for ways to make improvements and trying to spot errors in logic ahead of time. Hand tracing is when you play the part of the computer, recording each update to each variable, array, or hash as you walk through your algorithm. Force yourself to hand trace the algorithm all the way to the end, not just through the tricky parts. That way you might avoid a tiny, boring (but fatal) problem that just happened to be at the end.

4. From your point-of-view, as the reader of this book, it is possible that as a biologist in an interdisciplinary group, you might find yourself writing mostly in pseudocode to communicate with other group members who will actually write the lines of Perl.

Box 11.3 Importance of Beauty II
....

There is beauty in the tension between efficiency and understandability. You want to write code that makes optimal, efficient, and economical use of Perl's features; you don't want your program to look rigged up and convoluted like an intricate Rube Goldberg device. The path from A to B should not be full of detours. Code should be crisp and even a bit laconic.

Then in seeming contradiction to common advice, sometimes you do have to reinvent the wheel. Maybe Perl has what you need or maybe you are willing to stretch what you need to fit Perl. However, in some cases you may want to start from the beginning, tossing out all of the assumptions and rigging up your own little invention with duct tape and clothespins.

Meanwhile, you want your code to be absolutely transparent to future users, especially those who will need to add to, adjust, and otherwise maintain your program. That future user could be yourself, a year from now, puzzled by your own choices. Long descriptive variable names and essay-like comments take up both space and time. It takes longer to trace through code that occupies scores instead of dozens of lines.

To achieve that balance between efficiency and understandability is to produce something beautiful. Our friend and Perl guru Greg Williams, who like many gurus sometimes leans toward terse (rather than verbose), says that the goal is "to use as few of the language features of Perl as possible for the greatest effect while still being completely understandable."

This is something to which we all might aspire.

11.4 Jim Kent's BLAT Algorithm

A heroic story of algorithm building is that of Jim Kent, a programmer who was completing graduate work in biology at the University of California, Santa Cruz. When the sequencing of the chromosomes of the human genome had been completed, that genome was, at that point, far from finished. Like most freshly sequenced genomes, it was all in fragments, three million of them, described by Kent as being like the fragments of the Dead Sea scrolls. Assembly of those small DNA sequences was predicted to be a long and arduous process using the available software of the Human Genome Project. Meanwhile, the competing commercial enterprise at Celera was weeks away from assembling its own draft of the human genome and (at least for a time) keeping it proprietary as well as applying for patents on as many genes as possible.

Undaunted, Kent began sketching out algorithms (based on his research) by which the assembly might go more smoothly and quickly. Ambitiously, he focused on speed, requiring that the assembly be completed in about two weeks. Working as much as 80 hours a week out of his garage and using 100 Pentium III processors, Kent produced and implemented GigAssembler. It was used in June 2000, finishing the first draft of the human genome a few days earlier than Celera. Almost immediately, Kent and his group placed the draft into the public domain. In 2002 Kent published the algorithm for his tool, BLAT, which stands for BLAST-Like Alignment Tool. The goal of BLAT was to "stitch together" sequence fragments with significant overlaps. For example, these two fragments might be stitched together.

actgggca<u>ttagct</u>   and   <u>ttagct</u>atacgtgct

to make

`actgggcattagctatacgtgct`

and so on, to assemble an entire chromosome.

In his paper, Kent compared BLAST with BLAT, summarized and simplified here by us.

BLAST	BLAT
Query is some unknown sequence	Query is one of three million fragments
Database is all known sequences	Database is all of the other fragments
Builds index of short sequences in query	Builds index of short sequences in database
Scans linearly through database	Scans linearly through query sequence
Output shows areas of homology	Output shows stitched-together fragments between sequences

Both scan short matches and extend those to find the highest scoring match.

Kent noted that all fast alignment programs he knew of break the problem into two parts:

1. Search stage—Find all regions that are likely to be homologous.
2. Alignment stage—Examine regions from step 1 and make best alignments.

Presumably the search stage will screen out and reduce the amount of sequence passed on to the alignment stage, thus speeding up the more costly alignment.

Kent acknowledged that setting up the index of all possible short sequences is a hurdle but that once it is set up for a genome, the programmer "owns" it and the index does not have to be done again.

Box 11.4 Speed I

So Perl walks on water, right? Well, not always. One of the reasons that so many different programming languages have been invented and developed is because there are so many different types of problems. Languages often excel in some but not all situations. Here is a case in point.

One of us (the computer scientist) wanted to analyze 200 microbial genomes at once. The program in Perl was relatively easy to design and implement, but it ran slowly. Although significantly more difficult to implement, the program was rewritten in C++ in an attempt to decrease the run time for each experimental run. The experimental run times for 200 genomes was reduced from many hours to about one hour. For this unique case, C++ was a better choice, at least in regard to the time it took to complete an experimental run.

Similar to other languages you may know of (including Python and Java), Perl is not strictly a compiled language. So what does that mean? Every Perl program is processed by the perl interpreter. (Notice that the language Perl is spelled with an uppercase P and the interpreter or "engine" is spelled with a lowercase p.) The intepreter handles the execution of all Perl programs, much like word processing program (e.g., Word) handles written documents. The perl interpreter performs four steps: (1) loads a Perl program, (2) checks for syntax (typing) errors and assuming there are no problems (commonly called syntax "bugs"), (3) converts (compiles) each of the Perl instructions from the program to a middle-level language (a bytecode), and then (4) immediately "interprets" each bytecode to perform the requested action. The first three steps of the perl interpreter are not unlike those performed during the biological process of translation where ribosomes (the interpreter) read and convert codons to amino acids (bytecodes).

In contrast to the compile-interpret-run steps in Perl, languages like C++ are compiled and executed in two separate stages. C++ programs are passed through a compiler that checks for syntax errors, and then the compiler translates each C++ instruction into a series of very efficient low-level instructions, saving the low-level in a new executable file (e.g., pairwise.exe). This is the primary reason for the increase in speed in the previously mentioned C++ program: The compiler converts the C++ into low-level instructions that are perfectly suited for that computer. Note that compiling a C++ program does not execute the program. A second step is required where the user must run the compiled C++ program. On the opposite end of the language spectrum, pure-interpreted languages do not compile instructions into a low level or even translate to a middle level, but immediately interpret and execute each line of the program as soon as it is encountered. Purely interpreted languages are the slowest languages in regard to execution speed.

	Compiled language	*Compiled-Interpreted*	*Interpreted-only language*
High level	C, C++, Fortran	Perl, Python, Java	BASIC, JavaScript, shell scripts
example file name	pairwise.cpp	findRepeats.pl	moveFiles.sh
	↓	↓	↓
	COMPILER	COMPILER and INTERPRETER	INTERPRETER
	Check syntax, if okay then then create a new file containing the machine code	Check syntax in entire script, if okay, then compile into and then immediately execute the bytecode	Immediately execute each line of the script, halting if any line has an error
	↓		
Low level	machine code (pairwise.exe)		

11.5 A DNA Dictionary

Motivated by our collective pleasure when using the online *Oxford English Dictionary*, we decided to build a DNA Dictionary, or motif lexicon. We wanted to respond to a user's query (a short DNA motif of interest) with a rich set of annotation, including the type of motif (e.g., direct, mirror, inverted, or versatile repeat), all starting base pair (bp) locations of that motif within the genome, and links to information for upstream and downstream genes. Significant constraints included that a user could enter any motif of any length, and because the tool was Web-based, the search and response time had to be sufficiently fast. But how could we efficiently determine every location of any possible motif in an entire genome?

As the programmers in our group considered various strategies, the biologist in the group went to the whiteboard and shared her algorithm, "Search and save (cache) every starting location of every 4-mer, 5-mer, 6-mer, and 7-mer and concatenate the starting locations." (Insert silence from the programmers here.)

Here is how it was implemented. For every genome in the motif lexicon, we cached the starting locations of all 4 to 7-mers. An example of some starting locations for some of the 256 possible 4-mers is shown below.

Genome: AAAAACAAAATCG . . .
              1 2 3 4 5 6 7 8 9 0 . . .

Starting locations for each 4-mer:

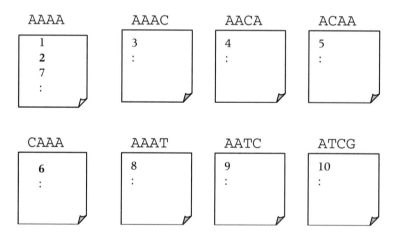

User Query: "Look up the motif—AAAACAAA "

Algorithm (using only 4-mers and liberties taken with variable names to match the example)

```
1) Get user query, e.g.: AAAACAAA
2) Break query into smaller 4-mers
          AAAA        CAAA

3) foreach  $AAAA_startBP (all locations of "AAAA")
   {
          foreach  CAAA_startBP  (all locations in "CAAA")
          {
                if  (AAAA_startBP + 4) == CAAA_startBP
                {
                      FOUND "AAAACAAA", record it.
                      Start again.
                }
          }

          } end foreach CAAA

   } end foreach AAAA
```

Of course, using 4–7-mers, query motifs could be partitioned into the fewest number of submotifs, for example, a 23-mer involved matching the staring and ending integer base pairs of

7mer—7mer—5mer—4mer.

No strings to search here. Caching and comparing the integer locations was the key to an elegant and efficient algorithm.

Box 11.5 Speed II

By speeding up the search, we slow down the user, encouraging contemplation and leisurely exploration. If the user can be sure that a dead-end query will not crank for minutes, then many more queries, motivated by nothing more than curiosity, will be tried. Curiosity is of course a foundation of science.

One problem we have noted in some sequence analysis software is that results are presented one at a time and so slowly that the user is likely to limit and edit his or her own queries to those most likely to yield an answer. Even worse, results may arrive by email, long after the query, thus discouraging repeated use.

In true exploration, even wrong paths may have surprises. Think of any seafaring explorer missing one island but finding another, unexpected and just as fascinating. Fast software allows you to dismantle and reinvent your hypotheses on the fly, making discoveries as you voyage along.

A necessary accompaniment to rapid searching is an absolutely clear results page, to remind you exactly what your search query and parameters were. This, too, will facilitate repeated use by creating a sort of map or log of where you've explored and where you might go next.

11.6 Breaking Sequence into Motifs

In chapter 4, we used a Perl program that applied regular expressions to a file of 7-mers, one motif per line. In this section, a highly annotated Perl subroutine (the final algorithmic phase) shows how you can chop a sequence of DNA into motifs of a user-defined length. Individual motifs are defined here as the motif within a window of length 7 bp. The next motif can be obtained by sliding the window 1 bp to the right, thus in this definition, motifs contain overlapping segments. The first two motifs to extract are shown next. The motifs are individually stored in the cells of an array.

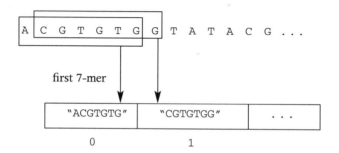

```
#----------------\
# breakIntoMotifs \
#------------------------------------------------------------------
# SUMMARY: Chop a sequence of DNA into individual motifs of a
# given length and store the individual motifs within cells of
# array. Return the array of motifs. Only saves motifs containing
# [ACGT]. Note: motifs are established by sliding a window
# through the sequence one bp at a time, thus adjacent motifs
# contain overlapping segments.
#
# IN: two arguments (1) DNA sequence ($text) and
#                    (2) length of motifs to extract ($LMER)
# RETURNS: an array of motifs each of length $LMER
#------------------------------------------------------------------
sub breakIntoMotifs
{
    my ($text, $LMER) = @_;        # shift twice

    my $DNAlen = length($text);    # get overall length of DNA
```

```perl
my @words; # array of motifs to fill and return when done

my $i = 0; # index of front of the window to slide down
            sequence
my $j = 0; # index into array @words to store motifs

my $end = $DNAlen-($LMER-1); # avoid sliding the window off
                             # the end
# while we keep sliding the window down to the next motif
while ($i < $end)
{
    # extract next motif from $text; starting at $i for $LMER
    # letters
    my $nextMotif = substr($text, $i, $LMER);

    # if the word contains a new letter (that was added to the
    # end) and that is NOT in [ACGTacgt], ignore this word and
    # skip downstream
    if ( $nextMotif =~ m/[^ACGT]/i )
    {
      # oh, this motif is not OK
      # must be an 'N' or some other nucleotide-code, so ignore
      # skip down to the next location of a possible valid word
      $i = $i + $LMER;
    }
    else                        # else this was a valid motif
    {
      $words[$j] = $nextMotif; # it is safe to save the word
      $i++;                     # slide window down
      $j++;                     # move array index down to next
                                # cell in @words

    }
} # while more words

 return @words;

} # end subroutine breakIntoMotifs
```

Box 11.6 Continued

smaller ones. To programmers, DNA is easily abstracted to a string, and strings are searchable. Once you've introduced yourself, try some of our favorite questions.

1. I've written a little section of a program that does _____, but is there more than one way to do this? The answer to that is so obvious that Perl programmers sometimes use an acronym for the triumphant answer: TIMTOWTDI, pronounced "Tim Toady", which means "There is more than one way to do it."
2. I want my program to do _____. Will I have to write that function myself, or do you know of a built-in Perl function I can use?
3. My program is a little sluggish. Could my algorithm be causing a bottle-neck somewhere? By the way, "sluggish" means it takes a second or two longer than instantaneous to return a result on a small test set. If that's what your program does with a short sequence of DNA, it is going to be much worse with an entire chromosome or genome.

In general, you, a novice in Perl, are going to be intriguing to your computer science colleague. Programmers love to show the capabilities of their favorite programming language, and they will enjoy your eager attention. Furthermore, most programmers like writing algorithms on the board, while drinking tall cups of caffeinated beverages; it is a sort of recreation. Being able to use some of the conventions in algorithm writing, described in this chapter, will help you bond quickly with other programmers.

12   Regular Expressions Revisited

In which some refinements will be introduced by which your regular expressions might be restrained from their customary greedy and exuberant activities. This is introductory obedience school for regular expressions.

Some people, when confronted with a problem, think: "I know,
I'll use regular expressions." Now they have two problems.

—Jamie Zawinski (in Conway, 2005)

Regular expressions (regex), introduced in chapters 4 and 6, deserve a little revisit. Before reading this chapter, you may want to review some aspects of regex syntax from those earlier chapters. Like an overgrown golden retriever puppy, regexes are lots of fun and eager to please. They are also destructive and a bit dangerous if trained incorrectly. Even professional programmers use regexes with caution.

Let's say a big, fluffy, golden retriever pup has been sent out to fetch the newspaper. Instead it fetches every single newspaper in the neighborhood, chews up some of them, leaves others in the bushes, and proudly brings the rest home. All that is to say that regexes are greedy in their eagerness to please. If you send one out to collect sequences of a particular length, you may expect to get the biggest sequence the regex could find, though you might have been hoping for something more subtle. This chapter shows you how to give your regexes some obedience training. If you're looking for a particular sort of sequence in a big file, what you probably want is the first and every subsequent occurrence of the pattern and not the single longest sequence the regex could retrieve for you.

Another challenge with long regular expressions is that they can be difficult to read if they are written out in one dense line of code. This chapter introduces the practice of splitting up a big regex into lots of little lines and commenting each section. You (and others who have to read your code) will not regret this in the long run. Regex syntax is not Perl syntax; indeed, it is far from it. It is like inserting lines of Greek or Chinese into English text. In the examples to follow, we make lavish use of comments to help the reader translate those regexes.

Regular expressions embedded in Perl rely on subtle syntax and some use what Larry Wall refers to as "magical Elvish variables," which is charming but adds to the difficulty of matching exactly what you were expecting. For example, global pattern matching variables introduced in chapters 4 and 6, such as $1 and $2, hold captured items from parentheses in the regex. Your results will be tucked into these special variables unless you use special syntax to prevent it. It is especially important to learn the syntax that enables noncapturing groups so that the regex does not become overloaded with secret caches of items. That's your golden retriever puppy again with lots of secret holes filling the yard with hidden treasure.

Intrigued? We hope so. In addition to warnings and recommendations, this chapter includes lots of refinements for your pattern matching search strategies and suggestions for how to use them in sequence analysis.

12.1 Commenting Regular Expressions

Regular expressions are sections of code written in a compact syntax. The subtleties of these pattern-matching instructions require documentation for the same reasons we comment the subroutines we write. For the same rationale for documenting sub-routines (e.g., see section 7.2), regular expressions require accompanying comments that help the reader translate your pattern-matching intentions into English. More often than not, the reader will be you, attempting to decipher your own regex.

The extended formatting mode (/x) allows you to comment your regular expressions. All whitespace (e.g., blank spaces and newlines) within your regex are not considered part of the pattern to match and are thereby ignored. In addition, the comment symbol (#) and anything following it to the end of the line is also ignored, thus you can comment each individual section of your regular expression like you do with other lines of Perl.

For example, the following regular expression is nontrivial to parse. What exactly is this intending to match?

```
if ( $motif =~ m/((.)(.)(.)(.*?)\4\3\2)/ )
```

Granted, we have thrown in one new feature that we have yet to introduce, but the point is that by using the extended formatting mode for regex (/x), this pattern-matching expression can be teased apart, translated to English, and presented to the reader in a very clear fashion.

```
my $dna = "ATCGTCATGCCTTCTATCGGTGCA";

if ($dna =~ m/
            (              # capture (remember) entire match
              (.)(.)(.)    # any three nucleotides, remember them, then
              (.*?)        # zero or more of any nucl, NON-GREEDY, then
              \4\3\2       # recall three nucleotides in mirror-order
            )              # end of entire match
          /x
    )
{      # if match is found
  my $mirrorRepeat = $1;    # save match with an appropriate name
  print "Found (non-greedy) MIRROR REPEAT (>= 6bp): $mirrorRepeat
  \n";
}
else
{   # else no match found
    print "No Mirror Repeat of minimum length 6 bp was found. \n";
}
```

Box 12.1 Go Ahead and Name It (and Join the DNA Philological Club)

Let's say you or your students are curious about ants and would like to immediately begin a project to find out which kinds of ants are attracted to a bait of peanut butter on the ground. A major stopping point is that you know nothing of ant taxonomy and would have to pore over dense dichotomous keys to have any hope of identifying a single ant to genus and species. That isn't a very gratifying start to an intriguing project. Tropical biologist Barbara Bentley suggests this solution. Get started with the peanut butter immediately. Then, as needed, make up your own temporary taxonomic key of whichever ants arrive. Maybe there are just three types. You go ahead and name them: "bighead," "tiny," and "red." Now you are free to make all sorts of fascinating observations about their interactions and activities, which is what you wanted to do in the first place. You could do the same with sequences, of course, well ahead of any serious search to see whether the sequence already has an official name. In your explorations, do not be held back at first by jargon invented by other people. That can come later when you are fitting your own research into the bigger picture.

Now let's say that you want to popularize your organisms (or your sequences), encouraging others to appreciate what you see and maybe join in. This is a realistic scenario for sequence analysis. In an interdisciplinary group, you need to cut through the jargon and demystify the naming process. A good example is what lovers of dragonflies did. They set out to popularize dragonflies, to bring them up to a similar status to that of birds or butterflies, by making up attractive common names for each one, For example, "goldenrings" instead of Cordulegastridae. The multisyllabic Latin name still stands as the most important identifier, but novices can enter the dragonfly club more easily. So it is with sequences. You don't want your first day in the interdisciplinary group to be spent stumbling over Latin jargon. Even if the terminology does not become popular outside of your lab, you have taken important first steps to building your own working community. One of the many examples from our lab is versatile repeat, coined by us for repeats that are both (inverted and direct) or are both (mirror and direct). We began using that phrase well before any serious analysis or deciphering of potential meaning.

Here is one pitfall of naming, particularly in a finalized form. The essential nature of and the act of naming and thereby sorting and classifying is sometimes reflected in premature names and categories—found later to be incorrect or impractical but which alas are fossilized in the literature. Biology is full of terms that must nonetheless be memorized if one is to follow the literature at any depth. Learning jargon is one of the most difficult tasks of undergraduate biologists, and it is further exacerbated by underestimates of just how important it is to be able to spell, pronounce, and accurately and spontaneously use multisyllabic jargon couched in two ancient languages (Greek and Latin). Furthermore, it is not always clear to a novice whether a difficult bit of

continued

Box 12.1 Continued

terminology is hard to learn of itself or whether because it is a relic, part of an
obsolete classification system with no apparent logic or connection to other
jargon. All this is to say that when it comes time to preserve and publish
(essentially fossilize) your newly coined name in the scientific literature, look
very carefully to see whether some preexisting name might work. We have
noticed that researchers of lower organisms are more likely to try to conform
to jargon because of the evolutionary connections between their favorite and
all others. On the other hand, work on human gene systems often acknowl-
edges, cites, and conforms to only terminology coined for humans. Better stan-
dards for coining official jargon will (perhaps someday) facilitate global
searching of databases for gene and protein names. Meanwhile a double or
triple vocabulary is sometimes necessary.

Finally, to end the topic of naming on a more positive note, the Unregistered
Word Committee is an inspiration. Its activities, documented in Winchester
(2003), are (as usual) relevant to DNA analysis. Isn't everything? In 1857, the
Unregistered Word Committee of the newly formed Philological Society
undertook a project that was to become the *Oxford English Dictionary* (*OED*).
They reasoned that the task of finishing the collection of all English words
might take one man 100 lifetimes, but might take 1,000 men just a few years.
Surely there were enough English-speaking scholars worldwide who would be
willing to volunteer their time toward a complete inventory of the English
language. A set of 54 pigeon-holes (each 6 × 6 inches) was built for the first
editor of the project. The holding capacity of the credenza: 60,000–100,000
half sheets of paper for as many word entries was deemed sufficient. However
when the project was actually completed, more than 68 years later, 6 million
equivalent half sheets contributed by thousands of volunteers had been
necessary. At the height of the project, 1,000 slips of paper were arriving to
the scriptorium per day. Today, the second edition of the *OED* (1989) contains
615,100 unique words defined using 59 million words. Truly the task of
compiling the *OED* is in the category of great public works, and the use of
volunteers was extraordinary. But then, so is the accumulation and annotation
of sequence data arriving daily to NCBI, one of the other great public works.
DNA (the barely readable book of life) is being deciphered and annotated one
"word" at a time mostly by volunteers, all part of a DNA Philological Club.
Welcome!

12.2 Greediness

Perl's regular expressions quantifiers are greedy, that is, the default behavior of the
regular expression quantifiers such as * (zero or more matches) and + (one or more
matches) is to find and return the longest match as long as the rest of the pattern
matches. This has significant implications when designing regular expressions that
contain patterns such as (. *) and (. +).

12.2.1 A Greedy Match

Given a long sequence of DNA, including an entire genome, how can we design a regex to find all the mirror repeats of length greater than 6 bp? The following example shows a first cut, but unfortunately the (.*) is greedy, and thus the pattern returns the longest match. In this case, greediness is not good.

```
my $dna = "ATCGTCATGCCTTCTCGTATCGGTGCA";

while ($dna =~ m/
            (              # capture (remember) entire match
            (.)(.)(.)      # any 3 nucleotides, remember, then
            (.*)           # GREEDY! 0 or more of any nucl.
            \4\3\2         # recall in mirror-order
            )              # end of entire match
            /xg            # find all (g) globally
          )
{ # while another match is found
        my $mirrorRepeat = $1;
        my $foundBP = pos($dna)-length($mirrorRepeat);
        print "Found (greedy) MR: $mirrorRepeat at bp
        $foundBP \n";

}# end while more MRs to find
```

The output of the regex with the greedy quanitifier (.*) is shown below.

```
Found (non-greedy) MR: CGTCATGC at bp 2
Found (non-greedy) MR: CGTATCGGTGC at bp 15
```

A careful look at the sequence will reveal that this regex has returned the second and longest mirror repeat (MR) and skipped past the initial MR.

What about this MR?

ATCGTCATGCCTTCTCGTATCGGTGCA

(.)(.)(.)   (.*)   \4\3\2
              greedy

Since the * quantifier is greedy, (.*) will match as much as it can. So would (.+). If you were to run this regex against a string representing an entire genome, you'd get back one very long result. In that case, you might suspect something was amiss, but imagine if you were searching relatively shorter sequences such as upstream regions in a prokaryotic genome. In this case, the MRs returned would look to be of reasonable length.

12.2.2 A Nongreedy Minimal Match (The Solution You Want)

You can force nongreedy minimal matching by placing a question mark (?) after your quantifiers. Thus (.*?) will minimally match zero or more and (.+?) will minimally match one or more. Adding ? after the quantifier in the regex will force the regex to match as few characters as possible, thereby finding the smallest MR that matches the rest of the regex.

```perl
my $dna = "ATCGTCATGCCTTCTCGTATCGGTGCA";

while ($dna =~ m/
            (                   # capture (remember) entire match
             (.)(.)(.)          # any 3 nucleotides, remember, then
             (.*? )             # NOT GREEDY! 0 or more minimally
             \4\3\2             # recall in mirror-order
            )                   # end of entire match
           /xg                  # find all (g) globally
         )
{ # while another match is found

     my $mirrorRepeat = $1;
     my $foundBP = pos($dna)-length($mirrorRepeat);
     print "Found (non-greedy) MR: $mirrorRepeat at bp
     $foundBP \n";

}# end while more MRs to find
```

The output when using the nongreedy quantifier is shown next.

```
Found (non-greedy) MR: CGTCATGC at bp 2
Found (non-greedy) MR: CGTATCGGTGC at bp 15
```

This time, the nongreedy quantifier finds both MRs. Perhaps you are thinking, "Gee, just one little symbol and such different results." Regexes are powerful, but with power comes responsibility.

MR at bp 2

ATCGTC**ATGC**CTTCT**CGT**ATCGG**TGC**A

MR at bp 15

Box 12.2 Hieroglyphics

Is a language (like English) with a linear (string-like) representation useful in generating hypotheses about information that might or might not be present in a DNA string? On first glance, an English-language (or linear-language) metaphor may seem like a reasonable first step at finding working hypotheses. Indeed, many of the examples in this book do just that: They treat DNA like a linear string of information that may be analyzed by some of the traditional methods of linguistic analysis, such as frequency analyses and index-building or concordance-building.

However, a purely linear approach has its limitations given the three-dimensional folds, kinks, and coils of DNA and the three-dimensional nature of DNA interactions with proteins and other molecules. Reluctant to let go of a natural-language metaphor prematurely and in the spirit of open-minded exploration, we suggest an examination of some of the less linear (more multi-dimensional) human languages especially in their written form. These might include Chinese ideograms and hieroglyphics.

Hieroglyphic ideograms typically are set out in linear fashion one after another, either right to left, left to right, up to down, or down to up. Like many ancient writing systems, hieroglyphics lack vowels and spaces between words. However, hieroglyphics do not have a simple one-to-one correspondence or transliteration with linearly represented languages, such as English or Russian. The hieroglyphic characters double back on themselves, three dimensionally, by containing layers of redundant information within each ideogram. That is quite a challenge for translators. Indeed, the code was impenetrable until the Rosetta Stone provided a translation in Greek for a string of hieroglyphics.

Middle Egyptian scribes had several choices as to how to write out a word in hieroglyphic notation. They could spell it letter by letter and leave it at that, but simple spelling is almost never done. There are 24 uniliteral letters in the alphabet, including *m*, which is a picture of an owl, and *w*, which is a quail chick.

The scribe could use diliteral or triliteral symbols so that for example *mw* becomes a new symbol, waves of water. Within a single ideogram, a scribe might choose to use both methods (the uniliteral letters and the diliteral symbol) to depict *mw* redundantly using all three—owl, quail chick, and water—for emphasis.

But that isn't all. That is merely the phonetic part of the ideogram. The owl sign also means "in" as a sort of derived pictogram and the waves of water sign may also mean "water." Thus, a third level of redundancy might be introduced. The pictorial representation for a particular word might be layered in redundantly as though to further emphasize the meaning.

To summarize, any given word in hieroglyphics is a complex, redundant combination of both its phonogram (how it ought to sound) and its pictogram (what it ought to mean). For example, the ideogram for *thirst* requires five symbols. (Keep in mind that vowels are left out.) The symbols are a reed (*y*) and a foot (*b*) representing the two consonants of the word, followed by a baby goat representing the diphthong for those two consonants, followed by the

continued

Box 12.2 Continued

pictogram of waves in water (water) and a crouching man holding his hand to his mouth (drink).

If this were done in English, then the word for *cough* might look like couFgh followed closely by one or more pictures depicting a person coughing. This would be the way to make sure that you could pronounce the *gh* like *F* and know the meaning from the picture.

Like any evolved system, hieroglyphics is convoluted in its lack of simplicity and shows many vestiges in its origins as a completely pictorial method of writing. The newer (5,000-year-old) phonetics seems to have been layered in cautiously leaving in extra cues to make sure of the meaning. Thus in hieroplyphics there are at least two methods of perception for deciphering a word: hearing and pronouncing the phonogram and seeing the pictogram. Finally, of course, context is important. One would hope to find the word *thirst* surrounded by other words that complete a logical sentence. Otherwise it is back to the dictionary for more deciphering as there are many opportunities for ambiguity, including abbreviations and transpositions of glyphs for reasons of aesthetics. Ironically, the great care to avoid misunderstandings by using layers of redundancy makes simple translation thousands of years later nearly impossible.

Like hieroglyphics, DNA can appear deceptively linear in its transliterated form, acggtgcca. The actual meaning is likely to be more complex, multilayered, and redundant, representing the likely kinking or coiling of the DNA at that site of the motif and the likely binding of one or more proteins. Thus the perception and meaning of a motif of DNA is the result of several types of chemical interactions, their probabilities of occurring and the context of the motif within the longer string. Furthermore, it is an evolved system and as such is expected to be a bit ambiguous and jerry-rigged with vestiges of former meanings.

Where does this leave the DNA analyzer? We submit with a deep respect for the ground truth of the lab bench. We cannot go back to ancient Egypt to clear up some of the many misunderstanding of hieroglyphics. We can, however, go into the wet lab or collaborate with colleagues who can do that for sequence interpretation. A certain motif may appear again and again in an enticing upstream position of a gene. It remains to go into the lab to find out what that motif actually does. How many different proteins can it bind and how securely? What is the shape of the DNA at that motif and under what conditions? What happens to the cell or organism if that motif is missing? It is just a list of hypothetical motifs without that information.

12.3 Noncapturing Groups

At times it is necessary to group a part of a pattern without capturing (remembering) the group in a hidden variable. Recall that the use of parentheses within a regex automatically captures (stores) the result of the pattern within the parentheses and the result of that particular match can be recalled either later within the same regex or after the regex is complete. For example, in the previous sections, we used \2 within the regex to recall the second captured item (the first nucleotide in a mirror repeat) and we used $1 in a subsequent Perl statement to recall the entire pattern that matched. Because grouping and capturing both use parentheses, there exists a special case that uses parentheses to group sections of the regex but does not capture the pattern that matches in that group. In certain cases, you need to use parentheses to group part of the regex together, but you do not need to recall the result of that part of the regex, so you want to avoid the overhead of storing the result. In other words, "match the pattern in this group, but don't remember the result." The syntax for using a noncapturing group is (**?:** *pattern*).

12.3.1 Is This an Open Reading Frame?

Assume you are scanning hundreds of segments of newly sequenced DNA strings and in preliminary analyses you ask the question, "Is this sequence an open reading frame?" More specifically, you would like to know if the sequence has a valid open reading frame within a start and stop codon. You are not necessarily interested at this time in the particular codons, that is, you do not need to store the start or stop codon nor the codons between the start and the stop. Thus, you do not need to capture any of your matches, you are just interested in whether a match occurs.

It is worth mentioning here that dealing with open reading frames in general and finding genes in particular is very complicated. To keep the focus on the noncapturing groups, this example makes a number of assumptions, such as ignoring the presence of exons. If you wish, imagine we are working with a prokaryotic sequence or are starting with cDNA.

To match an open reading frame but not capture the codons that make it up, a regex must match a start codon followed by any number of three-nucleotide codons followed by one of the three possible stop codons. Because we are only interested in the presence of a match and the sequence might be very large, capturing the codons would be a waste of memory. The following regex uses noncapturing parentheses to group the start and stop codons and all intermediate groups of three nucleotides.

A**ATG**GCTACAGAA**TGA**

A	**ATG**	GCT	ACA	GAA	**TGA**

```
# Non-Capturing Groups
#
# Is this an open reading frame? (Yes/No)

my $seq = "AATGGCTACAGAATGA";
```

```
if ($seq =~ m/
            (?: ATG)          # start codon (don't remember)
            (?: ...)          # 3-bp codon (don't remember)
            *?                # zero or more times, non-greedy
            (?: TAG|TGA|TAA)  # stop codon (don't remember)
         /x)
{
      print "Open reading frame found \n";
}
else
{
      print "No open reading frame \n";
}
```

The output when searching "AATGGCTACAGAATGA" is shown below.

```
Open reading frame found
```

Box 12.3 Braille, the Skytale Code, and Yet Another Metaphoric Connection
to Sequence Analysis

. . . .

In our quest to visualize DNA sequences (especially promoters) as something more than a two-dimensional line, we came on the metaphor of Braille as a writing form experienced through touch. In the analogy, the fingers are like binding proteins, coming into contact with the code of raised dots or promoter DNA sequences. Now take that linear presentation of information in Braille and wrap it, helix-like, around a stick. That would be a version of a Skytale Code, an ancient method by which a linear text written on a strip of paper is readable only if it is wrapped around a stick of the correct diameter. As for DNA, it is a helix, of course, albeit with super coils, bends, and kinks as well as a snarl of tangles all stuffed improbably into a nucleus. A simple smooth stick of regular diameter on which some Braille text is encoded goes only a short distance as a strong analogy. However there is a Skytale-like periodicity in DNA with 10–12 bases forming a complete turn in the helix. Other DNA contours most likely create other intriguing periodicities, although perhaps not so regularly. In sum, an aspect of sequence analysis, especially for understanding the binding of proteins, must entail an understanding of what physical contact the "fingers" of the protein can actually make with the "text." To gain an understanding of the physicality of the chemical information, check out some of the wonderful three-dimensional graphic representations of proteins at NCBI and other sites. It is worth loading the right viewing software for the pleasure of turning the protein models to any angle and to see their finger-like contact of proteins with DNA.

12.4 **Look-Ahead Assertions**

Look-ahead assertions allow the regex engine to peek ahead into the string without advancing the current position (pos). In the context of a while loop that repeatedly attempts to find all occurrences of a pattern within a string (using the /g global mode of the regex), each time the regex engine finds a match, the starting position for the next round of searching is set at the end of the previous match. At times, you need a way to know what's ahead but still hold back the current position of the last match. In previous examples where we (globally) find all matches, we compensate for the fact that the current position of the regex engine is located just after the last match by subtracting the length of the pattern from the current position after a match is found.

```
# the position after a match is at the end of the match

while ($dna =~ m/some regex pattern/g )
{
        my $match = $1; # save match
        my $foundBP = pos($dna)-length($match) ;
        print "Found $match at bp $foundBP \n";
}
```

In situations when you cannot afford to move beyond a certain point, the significance of the current position being at the end of a matching string after a successful match becomes important. This is not easy, so let's move to an example.

12.4.1 Motivation for Negative Look Ahead

Assume we are searching a DNA coding sequence and are looking for two specific consecutive codons: AAA (asparagine) followed by CGG (arginine). The following while loop repeatedly searches for and reports all patterns containing these two consecutive codons.

```
# find AAA followed by CGG (e.g., AAACGG )

$seq = "ACCAAACGGAAACGGCTCAAACGGTAG";

while ($seq =~ m/

                    AAA # Asparagine, followed by
                    CGG # Arginine
              /xg)
{
    my $foundBP = pos($seq)-6; # 6 matched bp
    :
}
```

Now try to restrict the pattern to only match if the two codons are *not* followed by one of the stop codons (TAG, TGA, TAA) It might be tempting to capture the codon following our match (. . .) and ignore any matches where the captured codon is a stop codon, as shown next.

```
# WRONG
# find AAACGG that is NOT followed by a stop codon

$seq = "ACCAAACGGAAACGGCTCAAACGGTAG";

while ($seq =~ m/
                 AAA      # Asparagine, followed by
                 CGG      # Arginine, followed by
                 ( ... ) # some codon, capture it
            /xg)
{
        # if captured codon does not match a stop codon
        if ( $1 !~ m/^(TAG|TGA|TAA)$/ )
        {
                my $foundBP = pos($seq)-9; # 9 matched bp
        }
}
```

Notice the use of the binary binding operator !~. Like the binding operator =~ that we have been using, !~ binds the string on the left-hand side to the pattern to match on the right-hand side, except in the case of !~ the return value is negated logically: "The captured codon does *not* match one of the following patterns". The if -statement is equivalent to

```
if ( not ($1 =~ m/^(TAG|TGA|TAA)$/ ) )
```

A problem with this overall solution is that in some circumstances, the code will fail to find all of the locations where AAACGG is *not* followed by a stop codon. The problem is that after finding a match, the loop comes back and starts looking for another immediately where it left off—*after* the captured codon. As shown in the following diagram, if the captured codon contained the start of another matching sequence (that is, AAA), the regex engine would overlook it and continue trying to match the pattern three base pairs farther down the sequence.

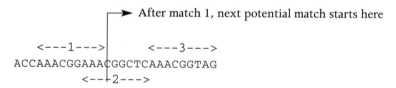

```
                              After match 1, next potential match starts here

         <---1--->            <---3--->
ACCAAACGGAAACGGCTCAAACGGTAG
             <--2--->
```

This sequence has three potential matches. Number one will match as expected. However, because the AAA at the end of number one was included in the initial match, the regex position has moved beyond that point. Thus, number two will fail to match. As expected, number three will fail to match because it is followed by a stop codon. The output confirms that we find only the first potential match. As feared, the second potential match is missed.

```
Found AAACGGAAA at bp 3
```

12.4.2 Negative Look Ahead

To match both 1 and 2, we need to use a "zero-width look ahead." A look-ahead assertion is a regular expression pattern that can tell you if a match will occur without moving the position in the string where the next match will take place. Technically speaking, we need to use a zero-width negative look-ahead assertion so that we can find a pattern that is *not* followed by a stop codon. The syntax for a negative look ahead is: (**?!***pattern*). Similar to the syntax for noncapturing groups in section 12.3, think of the question mark in this context as meaning "noncapture" and the bang (!) as "negative look ahead." So in this case, the parentheses will not capture anything.

```
# CORRECT
# Negative Look-Ahead
# Find AAACGG that is NOT followed by a stop codon

$seq = "ACCAAACGGAAACGGCTCAAACGGTAG";
#            3         9

while ($seq =~ m/
                AAA # Asparagine, followed by
                CGG # Arginine, followed by
                (?!TAG|TGA|TAA) # negative look-ahead
            /xg)
{
        my $foundBP = pos($problemSeq)-6;
        print "Found AAACGG not followed by STOP at bp
        $foundBP\n";
}
```

The correct output is shown below.

```
Found AAACGG not followed by STOP at bp 3
Found AAACGG not followed by STOP at bp 9
```

12.4.3 Positive Look Ahead

If we wanted to find all the locations where AAACGG actually *is* followed by a stop codon, we could use a zero-width *positive* look-ahead assertion. The syntax for a positive look-ahead is: **(?=***pattern***)**. Again, the parentheses will not capture anything.

```
# Positive Look-Ahead

while ($seq =~ m/
                AAA # Asparagine, followed by
                CGG # Arginine, followed by
                (?=TAG|TGA|TAA) # positive look-ahead
          /xg)
```

12.5 Summary of Regular Expression Syntax

Regex	*Meaning*
TATA	match four consecutive letters, TATA
TAG|TGA|TAA	match TAG or TGA or TAA
.	match any character but not a newline character
. .	match any two characters (independently, not necessarily the same character)
(.)	capture (remember) and match any character
.*	greedy match any character 0 or more times (each is independent of others)
(.*)	capture and greedy match any character 0 or more times
(.*?)	capture and nongreedy match any character 0 or more times
.+	greedy match any character 1 or more times (each is independent of others)
(.+)	capture and greedy match any character 1 or more times
(.+?)	capture and nongreedy, match any character 1 or more times

Regex	Meaning
\1	recall the first captured group
\2	recall the second captured group
\n	recall the *n*th captured group
.?	optional, match any character 0 or 1 time
T?	optional, match a T or nothing
(CAAT)?	Optional, match CAAT or nothing
A{3,7}	greedy match between 3 and 7 As
A{3,}	greedy match of 3 or more As
[CG]	match any *one* of the characters in the set, a C or a G
TATA[AT]	match TATA followed by an A or a T
[^CG]	match any *one* character that is *not* in the set, not a C and not a G
[CG]{5,10}	greedy match a C or a G between 5 and 10 times
^ATG	string begins with ATG
TAG$	string ends with TAG
(?:...)	cluster-only parentheses, don't capture 3 character match (don't remember 3 characters)
(?=TAG|TGA|TAA)	True if the look-ahead assertion succeeds; that is, it does find TAG or TGA or TAA
(?!TAG|TGA|TAA)	True if the look-ahead assertion fails; that is, it fails to find TAG or TGA or TAA
\s	match any whitespace character (tab, space, newline)
\S	match any character that is not whitespace
\d	match any character that is a digit, same as [0123456789]
\D	match any character that is not a digit
\w	match any one "word" character (includes alphanumeric, plus '_')
\W	match any one nonword character

13 Understanding Randomness

In which certain hazards of sequence analysis are noted, including attaching too much significance to discoveries and not being sufficiently cautious about "controlling" experiments for background noise with randomized datasets.

Indexes are . . . a place where the pleasure of sorting and ordering meets the opposite pleasure of the random, the inconsequential, and the chancy.

—Byatt (2001)

Most people do not have an intuitive understanding of randomness and, including ourselves, are susceptible to seeing and believing in the existence of patterns in random sequences. This can be a major challenge in DNA sequence analysis. Whether a pattern pops out to the naked eye or is revealed by pattern-seeking software, there is a great risk of misinterpreting the meaning of DNA sequences.

Our Intuition about Randomness Can Fool Us

Try this. Hand out two sheets of blank paper to each of several people. Their assignment is to write out on one sheet a sequence of 100 Hs and Ts, representing heads and tails of a coin flip, producing a sequence that looks as random as possible—random enough to fool you into thinking that it was not thoughtfully produced by hand, letter by letter. On the second sheet they are to record the results of 100 fair coin tosses. This will be another string of Hs and Ts, this time a result of mindless, random tumbles and falls.

Now go from person to person and glance at each of their two sequences. Try to guess which is the true random sequence (the one generated from 100 tosses) and which is the false random sequence (the one they created letter by letter). What is your trick for doing so? Look for the longest runs of Hs or Ts and most likely those will be in the true random sequence. Most people, in constructing a random-like sequence, cannot bring themselves to record a run as long as HHHHHHHH because it looks too much like a pattern and they know that patterns are not supposed to show up in random coin tosses. In fact, the reality is quite the opposite, and random sequences may be dangerously loaded with pattern-like strings.

Now try this. If you have acquaintances who regularly play a lottery game and who consider themselves fairly astute at choosing numbers wisely, ask them: which is a better number to play, 33333 or 43723? In many cases, the latter sequence will be chosen as the more appealing lottery choice because patterns like 33333 seem too meaningful and therefore too improbable. Of course, the chances of getting either sequence are identical. Next, ask the lottery players whether they would reuse today's winning number again the next day. Irrationally, the answer is often "no, of course not" on the grounds that the lucky number has had its win and is no longer due for further winnings in the near future. However, the winning number has no greater or lesser probability than any other for being drawn on the very next day. Disconcertingly, one can even ask these questions of students in a statistics class and discover that they have not lost any of their irrational instincts about choosing lottery numbers.

In many programs you might be writing for sequence analysis it is essential to run the same program on random strings of DNA to get a better picture of what is the potential significance of a sequence in your experimental string.

13.1 Using `rand` to Generate Random Sequences

A Perl function called `rand` can be rolled like a die to yield a pseudo-random fractional number from a specified range, for example, 0 to 3. This fractional number may be truncated to an integer, converted to represent a particular nucleotide or amino acid, and then concatenated onto the end of a string. Why pseudo-random and not random? A true random number is impossible to generate by any algorithm. Casti (1995) defines a string as random if there is no rule for generating it that uses a shorter string to write down the rule. Essentially, that means that a true random number is maximally complex. The Perl function `rand` does not have a nearly maximally complex method to seek a nearly perfect random number. Instead it has a sophisticated algorithm that is good enough for most purposes. However, if you generate an entire string of DNA with `rand`, one nucleotide at a time, it is best not to trust the first string you get as being sufficiently random. In good experimental fashion, plan to generate hundreds to thousands of random strings for comparison to get a better approximation. From this point, when we say or imply "random number generator," we mean pseudo-random.

`rand` (*EXPR*)

The `rand` function returns a random fractional number greater than or equal to zero and less than the value of *EXPR*. The value of *EXPR* should be positive and if omitted the value 1 is used, generating a fractional random number between zero and one, not including one. Because `rand` returns a fractional value, you should apply `int` to the returned value to convert the result to an integer. You may know that all pseudo-random number generators need to be seeded once before generating random numbers (i.e., primed to start). Perl has a seed function called `srand`, but `rand` calls `srand` automatically for you on the initial call to `rand`, so you do not need to explicitly call `srand` yourself.

```perl
#!/usr/bin/perl

use strict;
use warnings;

# FRACTIONAL values
my $n1 = rand( 1 );  # returns fraction: 0 <= value < 1
my $n2 = rand;        # same as above

# INTEGER values
my $roll = int( rand(6) ); # returns INTEGER: 0 <= value <= 5

my $nuc = int( rand(4) );  # returns INTEGER: 0 <= value <= 3

print "n1    = $n1   \n";
print "n2    = $n2   \n";
print "roll  = $roll \n";
print "nuc   = $nuc  \n";
```

One possible output of these four random values is shown below. Of course, if you run this program again, you will get a different set of numbers. Notice how int is required if you expect integer values. The int operator truncates fractional values, for example, int(2.9) becomes 2.

```
n1   = 0.514883738084794
n2   = 0.23388372487122
roll = 2
nuc  = 1
```

Also note that rand returns values up to but not including the value used as an argument. In the example of rolling a die, note that you must add one to your result because you expect numbers in the range of one to six, inclusive.

```perl
my $roll = int( rand(6) ) + 1; # simulate a die: 1 <= N <= 6
```

Try This Now

It is always a good idea to test your random number generator. An easy way to do this is to write a quick loop and print the randomly generated values. Run your program multiple times. You should see differing patterns of random numbers. For example, the following program prints ten randomly generated integers, each between 1 and 100.

```
# generate TEN random numbers: 1 <= N <= 100

print "10 random numbers: ";
foreach my $trial (1..10)
{
        my $N = int( rand(100) ) + 1;
        print "$N ";
}
print "\n\n";
```

The following output shows three different executions of this program.

10 random numbers: 87 99 90 66 97 66 54 20 52 70

10 random numbers: 95 11 75 63 6 51 67 46 89 9

10 random numbers: 85 55 68 81 32 43 56 35 9 9

....

Box 13.1 Ground Truth
....

We probably have not said enough in this book about the importance of molecular biology at the lab bench because of the primary focus on programming. The wet lab and actually the organisms themselves are the ground truth. Bioinformatics is a sort of remote sensing of a distant place, the genome. We cannot get there in person, and we still have a long way to go with a simple reading of a linear sequence.

One of the great advances in biology has been that there are so many different means of analysis that when they all converge on a bit of information we know we must be near or at a truth. Sequence analysis alone is fraught with false negatives and false positives. Finding a wonderful motif, say GTGACGTCAC, again and again with astonishing predictability in a set of promoters is just a single step in the remote sensing of genomes. Next must be the ground truth of the wet lab (not in the purview of this book), where the actual function (if any) of your favorite sequence might undergo some confirmation of its information.

13.2 Generating Random Sequences of DNA

You can use rand to help generate random nucleotides one at a time and concatenate each randomly generated nucleotide onto the end of a string. Ultimately, you want the randomly generated sequence to reflect the A:C:G:T ratios in your actual sequence of DNA. For example, a particular intergenic region of DNA might contain a high preponderance of As and Ts; thus, random sequences generated for comparison purposes should be more likely to contain a high number of As and Ts.

13.2.1 Random Sequences with Equally Likely Nucleotides (A = C = G = T)

Using rand to generate integers from one to four, inclusive, you could associate each of the four integers with a particular nucleotide, for example, A = 1, C = 2, G = 3, T = 4. In this case, note that we assume that the proportion of nucleotides in the sequence is 1:1:1:1. The following example generates 1,000 random nucleotides, counts the number of times we get each one, and prints the final totals.

```
# generate 1000 equally likely nucleotides

print "1000 random nucleotides: \n";

my $numA = 0;
my $numC = 0;
my $numG = 0;
my $numT = 0;

foreach my $trial (1..1000)
{
    my $N = int( rand(4) ) + 1;

    if ($N == 1) # A
    {
        $numA++;
    }
    elsif ($N == 2) # C
    {
        $numC++;
    }
    elsif ($N == 3) # G
    {
        $numG++;
    }
    else # must be a T
    {
        $numT++;
    }
} # end foreach trial
```

```
print " Number of As: $numA \n";
print " Number of Cs: $numC \n";
print " Number of Gs: $numG \n";
print " Number of Ts: $numT \n";
```

One sample output of 1,000 trials is shown here.

```
1000 random nucleotides:
  Number of As: 273
  Number of Cs: 254
  Number of Gs: 232
  Number of Ts: 241
```

13.2.2 Random A:C:G:T Distributions Based on Actual Sequence

The proportion of nucleotides in an actual experimental sequence is not likely to be 1:1:1:1. A few additional steps are necessary to get a pseudo-random string of the right proportion. Generating a random sequence of DNA of the same length and A:C:G:T ratios is something we are likely to want in a number of situations, thus, it is a great candidate for implementing this functionality in a subroutine. The following code assumes such a subroutine has been written. Following some Perl that calls the subroutine and a sample output, a complete subroutine is given.

```
# generate a random sequence of DNA based on the actual DNA

# start with an "actual" sequence of DNA
# Note: later on, we could read in the actual DNA from a file
# $sequence = readInDNA("someFile.fna");

my $sequence = "ACGGCTTGCCGTATATAAAATATACTATAA";
print "ACTUAL Sequence: $sequence \n";

my $randomSequence = generateRandomSequence( $sequence );
print "RANDOM Sequence: $randomSequence \n";
```

One possible randomly generated sequence is shown below. We leave it as an exercise for the reader to verify that multiple randomly generated sequences have a similar A:C:G:T ratio as the starting actual sequence. For now, notice that like the actual sequence, the random sequence also appears to be AT-rich.

```
ACTUAL Sequence: ACGGCTTGCCGTATATAAAATATACTATAA
RANDOM Sequence: TTACGAATACATTAGTAAAATTAACTATCA
```

The subroutine to generate a random sequence based on the actual sequence is shown here.

```
#----------------------\
# generateRandomSequence \
#------------------------------------------------------------
# Generate a random sequence of DNA with the same length and
# A:C:G:T ratios as the given sequence.
#
# IN: a sequence of DNA (assumes no newlines, sequence only)
#
# RETURNS: sequence of DNA of same nucleotide distribution and
#          length as the sequence sent in as an argument
sub generateRandomSequence
{
    my ($sequence) = @_;  # shift

    my $randomSequence;   # to hold the random sequence

    my $sequenceLength;   # the total number of nucleotides to search

    $sequenceLength = length( $sequence );
    if ($sequenceLength < 1)
    {
     print "Warning: sequence is empty; no sequence generated.\n";
     $randomSequence = "";
     return $randomSequence;
    }

    my $numA;        # will hold the number of each nucleotide
    my $numC;
    my $numG;
    my $numT;

    my $percentageOfA;    # to hold the percentages of each nucleotide
    my $percentageOfC;
    my $percentageOfG;
    my $percentageOfT;

    # convert to upper-case so we can ignore case
    $sequence = uc($sequence);

    # find and store the respective counts of the four nucleotides
    # Note: the tr utility can return counts of particular characters

    # find number of Adenine's (A), etc . . .
    $numA = ($sequence =~ tr/A/A/);
    $numC = ($sequence =~ tr/C/C/);
    $numG = ($sequence =~ tr/G/G/);
    $numT = ($sequence =~ tr/T/T/);
```

```
    # calculate the probability of each nucleotide here
    $percentageOfA = ($numA / $sequenceLength)*100;
    $percentageOfC = ($numC / $sequenceLength)*100;
    $percentageOfG = ($numG / $sequenceLength)*100;
    $percentageOfT = ($numT / $sequenceLength)*100;

    my $nextNuc; # hold new random nucleotide for each trial

    # generate a random sequence using the ACTUAL A:C:G:T ratios
    foreach my $trial (1..$sequenceLength)
    {
        my $N = int( rand(100) ) + 1; # random value 1 to 100

        if ( $N < $percentageOfA )
        {
             $nextNuc = "A";
        }
        elsif ( $N <= ($percentageOfA + $percentageOfC) )
        {
             $nextNuc = "C";
        }
        elsif ( $N <= ($percentageOfA + $percentageOfC + $percentageOfG) )
        {
             $nextNuc = "G";
        }
        elsif ( $N < = ($percentageOfA + $percentageOfC +
                          $percentageOfG + $percentageOfT) )
        {
             $nextNuc = "T";
        }
        else # trap cases with percentages of unknown nucleotides
        {
             $nextNuc = "N";
        }
        # concatenate new randomly generated sequence on to end
        $randomSequence = $randomSequence . $nextNuc;

    } # foreach nucleotide

    return $randomSequence;

} # end generateRandomSequence
```

Box 13.2 DNA Binding as a Fuzzy, Contextual, Combinatoric, Redundant, Evolved Language

....

In *Anatomy of Gene Regulation,* Panagiotis Tsonis provides "the most fre-
quent base–amino acid interactions" based on studies of 141 proteins. Thus we
have the beginning of a sort of list of rules by which regulatory proteins might

bind onto the promoter and enhancer sequences of DNA. If only we were not dealing with the probabilities and idiosyncrasies, of chemistry but rather with a clear one-to-one correspondence between protein structure and DNA sequence. Then we would be on our way to having enough information to directly and computationally decipher DNA binding sites, such as we do in reading triplet codons of genes. Instead we have percentages, representing the chance that a particular amino acid might find itself interacting with a particular base. Chip Lawrence (RPI) calls it "a secret world but badly spelled."

Guanine (G)
   22% with arg
   7% with lys
   7% with asn
   4% with gln
Adenine (A)
   17% with arg
   7% with asn
   6% with lys
   5% with gln
Thymine (T)
   12% with arg
   4% with gln
   1% with lys
   1% with asn
Cytosine (C)
   4% with arg
   2% with asn
   1% with gln

Additional information may be found in DNA repeats. Mirror and inverted repeats suggest that there might be a corresponding protein that is itself a mirror image or dimer. Direct repeats suggest a corresponding repetitive motif in a putative binding protein. The size (length) of a repeat might suggest the size of a corresponding protein or one of its domains.

Context is essential, although it is by no means an easy clue. In the tight space of a cell nucleus, DNA interactions are not necessarily linear. Motifs from thousands of bases away may be juxtaposed with a single loop. Furthermore, there must be a constant dichotomy between what is physically and chemically possible between DNA and proteins versus what is unambiguously useful and maintainable as information. A confounding issue is that some of the protein DNA binding is quite nonspecific, having much to do with the rather uniform phosphate-sugar backbone of the helix rather than the unique nucleotide sequence. Perhaps in the context of the cell, this makes sense because proteins are doing so many other things. The small subset of proteins used for transcription first needs to get into proximity of the DNA, regardless of its sequence.

Yet nonspecific binding must be an issue in the crowded nucleus. An analogy is typewriter keys, originally set out to be somewhat inefficient so that typing

continued

Box 13.2 Continued

fingers would not be so fast as to jam the key mechanism. If the fingers are protein domains contacting the DNA keyboard, it may be that well-spaced, specific, rare, and somewhat inefficient binding is what prevents chaos of contending proteins. Another analogy is the distribution of telephone area codes. Similar (and perhaps more memorable) numbers are not used within close proximity because a slip of the finger would result too easily in wrong numbers.

Thus, returning to DNA, the most confounding issue of all may be that the very best (quickest, firmest, most specific) interactions between protein and DNA are not necessarily the interactions occurring in a cell. Just barely good enough might be good enough and actually quite useful.

For another example of the usefulness of good enough (but not perfect), a temporal sequence of regulatory signals might be performed with a range of "spelling" from highly readable (or highly likely to bind a protein) to less readable (or more delayed in binding).

Skim this list of spellings for a well-known word. Note that consonants are more important than vowels, and this may be the case in DNA motifs as well—that some bases are more essential than others. These are essentially consensus sequences in the parlance of DNA analysis.

Elephant
Eliphant
Eliphint
Alefint
Alephant
.

.

.

Glepyeng

Except for the last entry, highly mutated especially of the consonants, the rest of the spelling variants are readable. Users of natural languages are of necessity quite capable of reading many spelling variations. Tolerance for diverse pronunciations and word orders is also high. Even faulty grammar is not necessarily a barrier to communication.

The redundancies of genetic information both in gene coding and in regulatory binding sites can seem puzzling at first. Removal of a motif (in the wet lab) does not necessarily shut down all functions. So is that motif unessential, or is it even functional at all? What are the limits of resolution in the wet lab assay that looks for function? The answer may be subtle. Consider the redundancies of the English (or any natural) language. There are many seemingly extraneous words that add subtle nuances of information. However the single word at the bottom of the list might be sufficient and perhaps the only truly essential motif in the sequence:

There is a fire in the wastebasket.
Fire in wastebasket.
Fire. Wastebasket.
Fire.

Furthermore, *fire* could be slightly misspelled or mispronounced, but the meaning would still come across.

So how do we know which motifs are being used in the expression of a particular gene? If we go in looking for a perfect or nearly perfect spelling, we may not find the motif. Though the elegance and perfection of good computer code is not a very helpful analogy for the evolved, jerry-rigged, ad hoc language of the genome, the problem of the vast quantity of information in a genome may be tractable by certain practices of computer programmers. Programmers often compete to write code in as few lines as possible and with the minimum information needed. Minimum information used in combinations can yield more complicated information. Thinking in terms of minimal code may give us clues about the lower limits of information needed for gene regulation. An upper limit may be the point at which the information slides around ambiguously, with several interpretations and a background of noise. Too many combinations may lead to contradictions.

13.3 Shuffling an Array to Get a Random Sequence

Another method to generate a random sequence uses an array and fewer lines of code to take any actual sequence and shuffle it like a deck of cards. This maintains the same proportions of As, Cs, Gs, and Ts. In this solution, individual nucleotides are first stored in cells of an array. Using a variant of an algorithm called the Fisher-Yates Shuffle, random pairs of nucleotides in an array are swapped over and over, shuffling the order.

The code to shuffle an array actually comes with Perl in a module. You just need to explicitly ask to use the module that contains the shuffle function (List::Util). Chapter 14 covers modules in some depth, but if you'll excuse the forward reference, the following example shows the shuffle function in action followed by a sample output.

```perl
use List::Util 'shuffle';

print "Using List::Util's shuffle... \n";
print "ACTUAL Sequence: $sequence \n";

# split on the empty pattern; returns individual letters
my @nucl = split ( //, $sequence );

# call shuffle and store into a new array
my @shuffledNuc = shuffle( @nucl );

# join individual letters into our new random sequence
$randomSequence = join ('', @shuffledNuc);

print "RANDOM Sequence: $randomSequence \n";
```

```
Using List::Util's shuffle ...
ACTUAL Sequence: ACGGCTTGCCGTATATAAAATATACTATAA
RANDOM Sequence: TAACAACTGGCTTTTTTAATGACAAAACAG
```

Box 13.3 Make Your Own DNA Die

A simple tetrahedron, with its sides labeled A, C, G, and T, makes a DNA die for generating (approximately) random DNA sequences. Templates may be found at educational software sites such as www.peda.com (look for their polyhedron templates).

Or copy, cut out, and paste this one from www.duke.edu/web/pfs/lessons/grade5math/Goal3/tetrahedronpattern.jpg.

TETRAHEDRON PATTERN

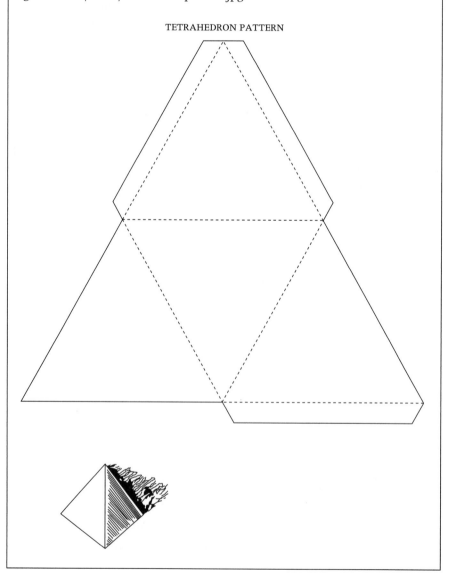

13.4 Weasel Programs: Interesting Applications for `rand` or `shuffle`

If you do an Internet search, you will find the Web site of writer and evolutionist Richard Dawkins (Richarddawkins.net), along with several links to computer programs that perform the weasel code as described by Dawkins (1996). The algorithm is briefly summarized in this section. The program demonstrates the power of cumulative selection as far superior to random searching when seeking a target word in a large search space. Cumulative selection is but one aspect (albeit an important one) of evolution that happens to lend itself to programming, but of course, it is by no means the entire story. Thus, the program is not a simulation of evolution per se, as some have argued in their criticism of it. An internet search for Dawkins and the weasel code will lead you to a range of commentary as people either over- or underestimate what the program is actually doing.

One example of a weasel program written in Perl is by Wesley Elsberry. You can find it (at Dawkins's site), cut and paste it into your IDE, and run the program. It allows you to choose some parameters, such as the mutation rate and the target string, such as "methinks." However, we particularly like versions using a particular DNA or protein sequence as a target. For example, you could start with a random DNA sequence and set the target to be a particular motif. Or you could start with a random amino acid sequence and set the target to be a sequence with the greatest number of hydrophobic amino acids (or amino acids of some other chemical property).

A simplified algorithm for such a program is as follows.

Create random string (of length L) of amino acids (or English alphabet or nucleotides).

Put the random string into two identical arrays (experimental arrays #1 and #2).

Choose a target sequence of length L.

Place the target sequence into an array (target array).

Begin While loop (while the chosen experimental array does not equal the target array)

  Mutate:
    Use random number generator to choose a number N within the length of the array.
    Use random number generator to choose an amino acid from the list of 20
        (or the alphabet or the nucleotides).
    Change one of the two experimental arrays by changing the Nth element
        in the array to the newly generated random amino acid.

  Compare and Select:
    Compare the two experimental arrays with the target sequence array, index by
        index.
    Choose the experimental array that best matches the target array.
    Discard the other. Replace it with replicate of best match

Count this as Generation #1 (#2, #3, etc).
    Repeat until experimental array is the same as the target

End While Loop

Report on the number of generations it took to get the match

....

Going Back For More

1. Write a subroutine to print the A:C:G:T ratios of a sequence of DNA. This would be a helpful subroutine to have available when you are trying to verify if your randomly generated sequences of DNA have "in the ballpark" ratios as compared to the starting actual sequence. Call the subroutine using both the actual and then each randomly generated sequence. See section 13.2.2 for additional details.

2. Implement a version of the weasel program.

3. Consider the following variant of using an array and a hash table to generate and hold the counts of randomly generated nucleotides. Type in and test this program. Include English comments to explain what each line of Perl is doing. After the `foreach` loop, what could you print to see the results?

```perl
my @nucleotides = ("A", "C", "G", "T");
my %nucleotide_count;

foreach my $trial (1..1000)
{
        my $n = int ( rand(4) );
        $nucleotide_count{"$nucleotides[$n]"}++;
}
```

....

14 : Modules

In which specialized packages of ready-made subroutines are
recommended and introductory instructions are given for finding,
loading, and using them.

The difference with Perl is that I decided to create the culture
at the same time as I was creating the language.
— Larry Wall (quoted in Davis, 1999)

We use modules every day, and therefore what is remarkable about them is easily overlooked. Nearly any complex task worth doing is made more efficient with modules. Say, I am building a new kitchen. Well actually I am not building it myself, I hire the module called "the plumbers," the module called "the electricians," and many other essential modules. Otherwise I have little hope of seeing my kitchen completed. Each of the modules I hire arrives not only with a repertoire of specialized expertise and time-honored practices but also with all of the necessary materials and tools. That same plumbers module is indeed modular; it can be used for any kitchen, and it is far more efficient than trying to learn and implement plumbing myself.

One of the things done especially well by computers is allowing one to divide big projects and use and reuse suites of functions to solve specialized parts of the project. Actually, nearly every aspect of programming introduced in this book (and in any programming book) has been about modularity. Code is typically written in little modular units that form an integrated whole. Even before writing the code, when you are still writing the algorithm on the board, the best practice is to break down the problem into modular parts. Furthermore, subroutines (introduced in chapter 7) make even more definitive the boundaries between sections of a program and the usefulness of calling the same specialized code again and again.

The actual word *module* has a particular meaning in computer science. It refers to any large package of related subroutines that may be loaded all at once and used, something like hiring a team of plumbers. Programming modules vary widely: Some are already loaded into the perl on your computer, and you have been using them all along. Some are relatively easy to load from the Internet and are directly useful for sequence analysis. Other modules are less user-friendly, especially for novices. In this chapter we merely introduce modules a good idea and a potentially functional part of your big project. The difficulty in writing this chapter is that many Perl modules are written and maintained on a volunteer basis by members of the Perl community. We cannot guarantee that a great-sounding module that you found on the Internet will be in perfect working order or that the instructions will work for your particular operating system. Therefore, this chapter limits its focus to a few modules that might be of interest in sequence analysis and that seem to have a history of being well maintained and relatively easy to use.

14.1 Using Modules Already Installed with Perl

Modules are collections of functions. Each of the functions within a module is very much like the built-in functions that we have already been using, for example, length, substr, and so on. Once you discover a function that you would like to use in a particular module, the first issue to settle is to find out if the module is already installed with your version of perl or if you will have to download the module from the Internet (see next section). This section assumes that the functions you want to use are within a module already installed with your version of perl. In this case, you need to request to use the module and specifically ask for the set of functions you need.

The simplest way of obtaining access to a function in a module is to simply use it, typically inserted at the top of the program, and also explicitly request (export) the functions within the module that you would like to use in your program.

```
#!/usr/bin/perl

use strict;
use warnings;

# load the List::Util module and export the max function
use List::Util('max');
```

The use operator takes a module name, finds the module on disk, and loads the code into your program, ready to be used. We recommend that you also export the function name(s) that you need at this time to both (1) show the reader which parts of the module you are using in this program, as well as to (2) allow you to use the short unqualified name (e.g., max) when you call this function. If you do not export the function name, you will have to use a fully qualified name (List::Util::max) when you make the call in your program. We show an example of the difference between exporting at the time of use and using a fully qualified name below. If the module cannot be found or loaded for some reason, the use operation will "die," and your program will quit with an error.

....

Try This Now

Because modules can vary greatly in their design and interface, the best way to learn how to use one and its associated functions is to read its documentation. With access to the Internet, the perldoc.perl.org Web site provides a searchable set of documentation. For example, search for List::Util. This will give you all the supplied documentation for the List::Util module, including some nice examples on how to use the functions supplied with the module. Notice in the example shown next, the documentation mentions that the List::Util module requires you to explicitly export each of the functions you want to use in this module. The qw (quote word) operator allows you to cleanly show items in a list; no quotes are needed around each item, and no commas are needed as separators between each item.

SYNOPSIS

```
use List::Util qw(first max maxstr min minstr reduce shuffle sum);
```

DESCRIPTION

List::Util contains a selection of subroutines that people have expressed would be nice to have in the perl core, but the usage would not really be high enough to warrant the use of a keyword, and the size so small such that being individual extensions would be wasteful.

By default List::Util does not export any subroutines. The subroutines defined are

Also, you can open a new terminal shell and use the perldoc command supplied with perl. On the command line, enter:

```
perldoc List::Util
```

If the module can't be found, you'll be given an error message.

. . . .

14.1.1 `List::Util`

The List::Util module provides a suite of functionality for working with lists. You could write most (if not all) of these functions as your own subroutines, but don't do that, for two reasons. First, these functions have stood a test of time and were deemed correct and valuable by the larger Perl community. Second, these functions are implemented to be very efficient.

For example, in List::Util, a max function will return the largest value in an array of values, where largest is defined by the greater than numerical comparison operator (>). If you do not export the function, it may be called with its fully qualified name, as shown below.

```
use strict;
use warnings;

use List::Util; # use module, no functions exported

my @redSignals = (1500, 1246, 1099, 2798, 1003);

my $largest;

# must call the function in the module using
# the fully qualified name

$largest = List::Util::max( @redSignals );

print "The largest red signal is: $largest \n";
```

```
The largest red signal is:  2798
```

As mentioned earlier, we prefer to export the functions with the use operator.

```perl
use List::Util('max');       # use module, export max
my @redSignals = (1500, 1246, 1099, 2798, 1003);
my $largest;
$largest = max( @redSignals );
print "The largest red signal is: $largest \n";
```

Often you will need to use more than one of the functions in a module. In this case, we recommend that you export them all at once. Note that because function names must be quoted, we recommend that you use the quote words (qw) construct to quote a list of space-separated words.

```perl
use List::Util qw(max min shuffle);
```

14.1.2 **File::Glob**

As discussed in section 9.4, the default glob function (CORE::glob) may not handle directory and file names containing whitespace. Thus, if you try to glob some files in a directory that has whitespace in one or more parts (e.g., if your file name was intergenic krebs cycle.fna or on Windows computers, if your path name included Documents and Settings), glob may not work. The module File::Glob to the rescue! This module includes a glob function that does not split on whitespace and thus can handle directory and file names with whitespace as shown in the next example.

```perl
use File::Glob('glob');       # export the glob function
my $fileList;
# DNA files held in directory: "bacteria and archaea"
$fileList = glob("bacteria and archaea/*.fna");
```

Box 14.1 What About BioPerl?

....

BioPerl comes up often on Internet searches for applications of Perl in the biological sciences. The first couple of pages of BioPerl (www.bioperl.org) show it to be comprehensive (even exhaustive) and well populated by some of the heavy hitters in computational biology. A search of your favorite keyword at BioPerl is likely to reach an intriguing page that recommends and describes an appropriate Perl module.

So your logical question is: Why are we not covering BioPerl modules in this chapter about using modules?

The answer is that after much scrutiny and discussion, we consider the modules at BioPerl to be an advanced topic, somewhat out of the scope of this book. Many of the BioPerl modules are in fact modules within modules. Some can be downloaded from the BioPerl site; others must be found and downloaded elsewhere. The complexity and size of the BioPerl modules are such that the downloads are nontrivial. We could tell you to get them and load them, but we really should include many pages of detailed instructions. In short, although we hate to phrase it in discouraging terms, many of the BioPerl modules are "expert" modules. We prefer to keep your first experiences with Perl as completely gratifying and uplifting as possible, and that just may not be doable with BioPerl.

However, all is not doom and gloom. You may well be one of the many scientists working in an interdisciplinary group. Your contribution might be not to download a BioPerl module but to point out their existence and suggest that a particular one might be further explored by someone who feels comfortable with the command line and a long list of queries in command line jargon. Then, once it is up and running (and tested), step in and try it!

14.2 Introducing the CPAN

In Perl, it's often the case that some bit of functionality you need for a program has already been written and packaged into a module. But where can you find it? One of the biggest advantages of using Perl is the Comprehensive Perl Archive Network (CPAN; www.cpan.org). The CPAN is a worldwide archive of Perl modules available for free and accessible online.

To demonstrate the power of CPAN, we walk you through the process of downloading and installing a module. Later on, we'll see how to search the CPAN for modules that perform specific tasks.

Let's assume we have need of a statistics module for performing standard deviation and other statistical tests. There are many modules on the CPAN that provide this functionality, each with a slightly different interface and complexity level. For our purposes, the `Statistics::Lite` module is an ideal choice given its simple interface and the fact that it has no external dependencies on other modules.

14.2.1 Installing New Modules from CPAN

Instructions for installing modules from CPAN are listed on their Web site. Follow the instructions for installing a new module to install the `Statistics::Lite` module.

14.2.2 Using New Modules Installed from CPAN

To use a newly installed module, follow the examples provided in the module's documentation. In the documentation, the Synopsis section should give you a brief introduction to using the module in the form of a script excerpt.

The following example assumes you want to use the `Statistics::Lite` module. Because this module does not come preinstalled with perl, you must download and install it first.

In the case of `Statistics::Lite`, you must first ask your Perl program to load the module and export the group of functions you need.

```
use Statistics::Lite (':all');
```

Here, the (`:all`) tells the `Statistics::Lite` module that you'd like access to the entire group of its functions. If you know you'll only use a subset of the functions, you can export other individual function names, such as

```
use Statistics::Lite qw(mean median mode);
```

See the Import Tags section of the documentation for more information.

Now that the module is loaded and all the functions have been exported, you may access all the functions described in the documentation.

```
use Statistics::Lite (':all');
   :

   :
# assuming @data is now filled with values

# statshash returns a hash whose keys are the names of all
# the statistics functions in the module with the
# corresponding values as the answer(s) for that function.

my %data = statshash @data;

# fetch the result for the stddev of all the @data
my $dev = $data{'stddev'};

print "standard deviation: $dev\n";
```

Box 14.2 Fonts for DNA
. . . .

What if there was a module (perhaps a collaboration of a Perl programmer and a type designer) that improved the readability of a DNA sequence, perhaps with a range of different fonts or symbols for different types of analyses?

The letters A, T, C, and G are far from ideal as graphic representations of DNA. The use of those letters is merely serendipitous, a consequence of the initial letters of the chemical names It strains the eyes to scan DNA sequences, trying to count bases or spot a particular motif or triplet and it is not necessarily our fault.

ATTCGTATCACGTTAGGGCATATCAAAAGTCTCTTTGTTTACGTACGCATACG

The Cs and Gs are essentially identical. The thin single leg of the T produces disproportionate white space in the line. Depending on what letters it is next to, the A sometimes appears to be hiding and other times to form a mountainous landscape. What would be better? Almost anything.

We think that there is considerable room for a clever font designer to explore some other possibilities. Simply trying out the range of fonts available on your word processor is a start. Note that included with your font collection is probably at least one set of symbols, such as Zapf Dingbats by prolific type designer Hermann Zapf. It may be that symbols would be a better system. Or perhaps a different alphabet would work, such as the ancient Celtic system of Oghams especially used to write on sticks.

A couple of examples are presented in *The Pattern Book* edited by Clifford Pickover. Ulrich Melcher suggests a Braille-like representation for DNA that deliberately emphasizes and sharpens the differences between the four symbols. Thomas Schneider presents "sequence logos" by which letters of various heights symbolize the probability of a particular base appearing in a sequence.

All this is to say that we do not have to settle for the usual sequence presentations of nucleotides. It might be an interesting interdisciplinary project to invent something better and then present it as a module.

14.3 Searching the CPAN

The easiest way to find what modules are available on the CPAN is to use the CPAN Search Site, located at search.cpan.org. Some of the modules are categorized and available from links on the main page: Data and Data Types, File Handle Input/Output, World Wide Web. A more direct approach is to use the search field to find modules that match a specific keyword or words.

14.3.1 Searching for `WWW::Search::PubMed`

As an example of searching for modules on CPAN, go to the CPAN Search Site and search for PubMed. This will search the documentation for modules on CPAN, returning a list of matching modules, documentation, and module authors. Among the search results will be the `WWW::Search::PubMed` module, which provides access to search results from the National Library of Medicine's PubMed database.

The CPAN Search Site provides a convenient way to view module documentation. By clicking on the name of a module on the search results page, you'll be able to directly view the documentation for that module. Module documentation typically follows a standard layout with common sections, such as Synopsis, Description, and Functions. By glancing over these sections, you have the chance to make sure the code looks like it does what you expect before installing the module. If the module seems to satisfy your needs, the next step is to download and install it using the instructions on the Web site.

14.3.2 Using `WWW::Search::PubMed`

The `WWW::Search` modules provide a way of accessing results from a variety of search engines. The `WWW::Search::PubMed` module allows your program to search the NCBI PubMed abstract database. Looking at the CPAN Search site for `WWW::Search::PubMed`, we can see that to install the module, we also need to install two other prerequisite modules: `WWW::Search` and `XML::DOM`. Install these modules and try out a variation of the example code from the Synopsis section of the module's documentation.

```
use WWW::Search;

my $s = new WWW::Search ('PubMed');

$s->native_query( 'ACGT' );
my $r = $s->next_result;

print "Title: " . $r->title . "\n";
print "Description: " . $r->description . "\n";
print "URL: " . $r->url . "\n";
```

This code will print out the first result from a PubMed search for the ACGT motif. A result is shown below.

```
Title: [Nucleotide correspondence between protein-coding
   sequences of Helicobacter pylori 26695 and J99 strains]
Description: (Mol Biol (Mosk). 39(6):945-51)
URL: http://www.ncbi.nlm.nih.gov:80/entrez/query.fcgi?cmd= Retrieve&db=
   PubMed&list_uids = 16358730&dopt=Abstract
```

This is a powerful module! It allows your Perl programs to query PubMed abstracts automatically, saving the results to parse, collect, and analyze. We trust the implications for data mining are obvious and exciting.

We acknowledge that this example uses syntax that hasn't been covered in this book. In this object-oriented example, $s and $r are objects, and next_result, title, description, and url are pieces of code (called methods) that can be called using the object as an implicit argument. For now, trust that the calls to the methods are similar to calling functions and you can use the results in similar ways.

14.3.2 Using LWP::Simple

LWP is a module used for retrieving data from various types of Internet servers: Web servers and FTP servers being two of the most common. LWP stands for libwww-perl, a collection of modules that provides functionality when working on the Web. You should install LWP using the instructions from the Web site (see section 14.2.1). Note that the LWP module may prompt you for configuration options during installation. It is generally safe to use the default answers to all of the options.

To show how LWP can be used, we use it to download bacterial DNA files from NCBI. We use the *E. coli* genome (K12 strain) as an example. The NCBI FTP address for *E. coli* K12 files is:

ftp://ftp.ncbi.nih.gov/genomes/Bacteria/Escherichia_coli_K12/

The LWP::Simple module provides a simplified interface to the full features of LWP. The simplest way to use the features of LWP is to use the get function provided by LWP::Simple as shown next.

```
use LWP::Simple ('get');

my $fileList;

$fileList =
    get('ftp://ftp.ncbi.nih.gov/genomes/Bacteria/
    Escherichia_coli_K12/');

if ( !defined($fileList) )
{
        die "Undefined value returned from LWP::Simple::get
        function \n";
}
```

The get function will put an FTP file listing into the $fileList variable. To parse the file listing to extract the file names and decide on which ones we want to download, we use a regular expression. The file list includes lots of information, but only the last field concerns us: the file name.

```
-r--r--r--  1  ftp   anonymous  33039681  Dec   4  13:10  NC_000913.asn
-r--r--r--  1  ftp   anonymous   1750995  Dec   4  13:10  NC_000913.faa
-r--r--r--  1  ftp   anonymous   4267327  Dec   4  13:10  NC_000913.ffn
-r--r--r--  1  ftp   anonymous   4706025  Dec   4  13:10  NC_000913.fna
-r--r--r--  1  ftp   anonymous     58517  Dec   4  13:10  NC_000913.frn
-r--r--r--  1  ftp   anonymous  11940963  Dec   4  13:10  NC_000913.gbk
-r--r--r--  1  ftp   anonymous     82569  Oct  19   2001  NC_000913.gene2
-r--r--r--  1  ftp   anonymous   7101533  Dec   4  13:10  NC_000913.gff
-r--r--r--  1  ftp   anonymous    366742  Dec   4  13:10  NC_000913.ptt
-r--r--r--  1  ftp   anonymous     10236  Dec   4  13:10  NC_000913.rnt
-r--r--r--  1  ftp   anonymous       232  Dec   4  13:10  NC_000913.rpt
-r--r--r--  1  ftp   anonymous  11922776  Dec   4  13:10  NC_000913.val
```

If we want to download only the two files of DNA sequence and protein table, we're only concerned with files ending in .fna and .ptt. When we've found the files we're interested in, we use LWP::Simple's getstore function to download the files and save the entire contents of those files to local files.

```perl
my $rc; # to catch the HTTP response code status

while ($fileList =~ m/([.\w]+)$/gmsx)
    {
        my $filename = $1;
        if ($filename =~ m/(fna|ptt)$/)
        {
                print "$url$filename";
                $rc = getstore( "$url$filename", $filename );
                if ( is_error($rc) )
                {
                        die "getstore failed; status = $rc \n";
                }
        }
    }
```

Running this code will result in two files, NC_000913.fna and NC_000913.ptt, being saved in the directory where we are running the Perl program.

....

Going Back for More

1. Look up the documentation for the reduce function in the module List::Util. What does this function provide? Can you think of an example where you might want to use it? Write a small program to test a use of the List::Util::reduce function.

2. Get a mug of coffee. Browse the CPAN Web site. Take your time and enjoy.
3. Install the `Statistics::Lite` module. Write a small program to test some of the statistical functions.
4. Practice data mining. Install the `LWP::Simple` module and search PubMed for some of your favorite abstracts. How could you collect information found in the abstracts that lead to further searches?
5. Write a Perl script to search for PubMed abstracts that relate to your own research. Parse the results for the information that most interests you. Craft a nice report.

Box 14.3 Musical DNA

What if there was a module that allowed you to input any sequence and then played it back to you as distinctive sounds, either music-like or not. That way, in addition to visual analyses, you could take advantage of the fine auditory discrimination that allows you to recognize voices, pick out themes in fugues, or identify syncopations when they occur. The ear may be an underutilized sense for examining sequences.

Furthermore, the important information in DNA sequences may not be pixel-like or quantum-like. Although we collect, store, and analyze discrete sequences, the real message may be more of a flow, like a video or like music. If we click through a video frame by frame or through music note by note, we miss much of the temporal and spatial information.

Many Web sites feature music composed by using combinations of DNA bases or amino acids as notes. Try an Internet search of "DNA music" or "protein music" to listen to many fascinating musical interpretations of particular genes or proteins. The primary reason for converting biological sequences to music seems to be an aesthetic one so far.

We are not aware of any attempts to identify sequence patterns with sound, but it might be a creative approach worth trying. It could be that using four very distinctive (not necessarily musical) sounds might allow the ear to pick up certain patterns that the eyes do not. Or perhaps a more derived, complex, and musical rendering of a sequence would reveal more complex fugue-like patterns.

This is a potential module that might be an enjoyable collaboration for a musician or composer or sound engineer with a Perl programmer.

15 Conclusions

To travel hopefully is a better thing than to arrive.

—Robert Louis Stevenson, "El Dorado" in
Virginibus Puerisque (1881)

Anton van Leeuwenhoek, the microscopist, was in great demand as a party guest in fashionable Delft homes in the seventeenth century. He could be counted on to bring along several handheld microscopes with glass bead lenses on which he would mount diverse specimens, including scrapings from the teeth of his fellow guests. A contemporary of van Leeuwenhoek, essayist Constantijn Huygens, wrote about such party scenes, noting the difficulty of people seeing for the first time the teeming microscopic inhabitants of their mouths. Without previous experience and a context in which to classify the new information, most complained that they had seen nothing: "When even completely inexperienced people look at things they have never seen, they complain at first that they see nothing, but soon cry out that they perceive marvelous objects with their eyes" (Alpers, 1984).

Often we simply do not see what we do not (yet) understand. The problem in sequence analysis is compounded by the sheer quantity of the unanalyzed information and the indirect methods by which that information is deciphered.

How do we build tools to seek out particular types of sequences when we do not know the full extent of what might constitute a meaningful sequence? Phrasing a question and building a computational tool to answer it are activities in forming hypotheses. They are worthy scientific enterprises in and of themselves.

What do we do with the results of genome analyses? These, too, are potential hypotheses for continued searching, for tweaking tools, and designing new ones. Finally there is the ground truth of the molecular biology wet lab where sequence results can provide the hypotheses for experiments.

How should results of sequence analysis be organized? It should be done with an appreciation that any given organizational system is just one of many choices. Furthermore, we are often limited to visualizing data in only two to three dimensions. Some multidimensional correlations may be invisible depending on our choices. Deciding on a particular presentation of data is also a form of hypotheses building. It is important to be ready to take advantage of serendipitous correlations, juxtapositions, and relationships that tumble out of any large data set. Keep the hypotheses flexible and know when to abandon one. Fortunately, some programming approaches let you do that fast, like abandoning a dead-end Internet search.

How much information is slipping through our current methods of classification and decoding because we do not yet have the context by which to design the right search tools? Industrial labs, academic labs, granting agencies, and editors of journals often have clear visions and definitive opinions of what sorts of sequences are worth collecting, analyzing, and using; traditionally, gene sequences have been featured high on their lists. Indeed the annotation of any new genome begins and often ends with a catalog of which genes are present. Often the sequences between genes are ignored. When such intergenic sequences are analyzed, they do not become neatly classified and organized into preexisting databases designed with genes in mind. Meaningful information, which often reveals itself in well-organized data sets, may not be visible at all. That means there is plenty of work for anyone, even single researchers or small lab groups, to take their own slice of the pie, defined as they wish.

Expeditions of free exploration and discovery are what drew many of us into the sciences when we were children. As depicted in *National Geographic* and *Nova*, scientists we admired seemed to be always on the go and on the verge of a great unknown. That enticing, exhilarating idea of being on an expedition is (or could be) an aspect of DNA sequence analysis, especially in its current status. The balance is tipped heavily toward vast, unknown territories of undeciphered data waiting to be explored.

Appendix 1

How to Isolate DNA from Strawberries

This is from the classic method for DNA isolation, for example, it was used by Franklin, Watson, and Crick when they isolated DNA for their X-ray analyses. Strawberries are an especially pleasant starting material. (Other procedures recommend beginning with a piece of a thymus gland from a calf.) As a stringy, slimy molecule, DNA readily comes out of solution and displays itself as a gelatinous mass. This procedure is so easy that one of us (the biologist, of course) surreptitiously isolated some DNA (as a sort of demonstration for a computer science colleague) in a diner in downtown Worcester, Massachusetts, while eating breakfast.

1. Cut up two strawberries and place them in a sealable plastic sandwich bag.
2. Add about 100 ml (3/8 cup) of chilled extraction buffer (made of household ingredients):
   Extraction Buffer
   (keep in refrigerator) (enough for about three extractions):
   1/8 cup (33 ml) of simple shampoo (without conditioner)
   2/3 tsp (5 g) salt
   1 1/8 cup (300 ml) water
   (The shampoo contains a detergent, sodium lauryl sulfate, that breaks apart cell membranes, releasing the DNA and other cell contents. The salt helps make a solution that is similar in saltiness to the inside of cells.)
3. Seal the bag and squish the strawberries, being careful not to squirt the mixture from the bag.
4. Pour the strawberry mixture through cheesecloth to remove the pulp.
5. Pour the red liquid into a tall, narrow, clear vessel.
6. Slowly pour about 50 ml (about 1/4 cup) of ice-cold ethanol down the side of the vessel to form a layer on top of the red liquid. (Keep the ethanol in the freezer until you are ready to use it.)
   For best results, use at least 90% alcohol. In some states, that can be purchased (by people over age 21) as grain alcohol. Alternatively, you could ask a chemistry or biology teacher for 50 ml of *denatured* ethanol. It will contain some methanol, rendering it Poisonous to drink. (That's what "denatured" means in this context.) High-proof alcohols like vodka are actually not concentrated enough for optimal results. For example 150-proof vodka is only 75 percent alcohol.
7. Notice the filmy white DNA precipitating where the ethanol and red liquid are interfaced.

8. Use a glass rod to gently spin some DNA from the interface. (Alternatively you could use some kitchen item like a toothpick, however the use of the glass rod is classic.)

9. Note the thready, mucousy nature of the DNA. If you touch the glob to the side of the glass and gently pull it away, you may catch a glimpse of a gossamer thread forming temporarily.

10. When done, rinse all solutions down the sink and rinse glassware.

How About Other Starting Materials Besides Strawberries?

Strawberries work well because they can be macerated by hand in a plastic bag. Bananas and kiwi fruit work, too. For tougher fruits and vegetables or fresh (or frozen) meat, you need to chop finely and then run through a blender. If you do choose to try animal tissue (meat), the best choice is organ meat like liver or kidney or sweetbread (thymus), as fresh as possible or frozen.

Source: Based on a procedure by Diane Sweeney, Pearson Education (carnegieinstitute.org).

Appendix 2

Bioinformatics as Part of the Solution

Jane Margolis and Allan Fisher asked the difficult question of why there are so few women in computer science. The results of their study, focused mostly on the computer science department of Carnegie Mellon University are presented in *Unlocking the Clubhouse* (2002). The authors included a list of suggestions for enriching computer science assignments so that they might be more broadly appealing to women, who (according to the study) are less attracted to more conventional programming problems. (We would add that a minority of men are attracted to computer science and these same suggestions might broaden male participation in computer science as well.) We noticed in reading the list of suggested enrichments that bioinformatics may be part of the solution. Indeed, our own approach to Perl has been primarily centered on the joys of applications and the intriguing personalities and problems of bioinformatics. Therefore we are reprinting the list of Margolis and Fisher, shortened by us and including our own commentary (in parentheses) on the excellent fit of Perl and DNA sequence analysis.

Ways to Enrich a Programming Assignment (or Career in Computer Science)

1. Make it useful, personal, or local. Make it socially relevant. Include observations of the real world. (That's bioinformatics! All sorts of fascinating, practical problems in medicine, the environment, and evolution may be solved through computation.)
2. Interface with other programs. (The public repository of sequences at NCBI includes powerful programs for analysis such as BLAST. We encourage a balance of using what is already available and writing your own as needed.)
3. Focus on the ease of use. (Throughout this book we emphasized how do-able this is without necessarily being a wizard or guru.)
4. Use big data. Use real-world data. Problems that are too large to solve by hand help students understand the real value of computing. (Are billions of bases of DNA a big enough data set? The sequence files are growing every day. There is enough unanalyzed information for lifetimes of work. We are in an age of genome exploration.)
5. Use natural-language text. (We were pleased to see this one because our whole approach to sequence analysis is to treat the data like a text to be deciphered. Our working metaphor is linguistics.)

6. Make it sensory: graphics, audio, animations. Write simulations. (Although we don't go there in our book, graphics and simulations are an obvious next step for creatively presenting the intricacies of DNA.)

7. Bring in experts. (One of our favorite teaching methods is linking preexisting courses in which, for example, a biologist does a couple of lectures in computer science and vice versa and then the two sets of students collaborate on an interdisciplinary project in DNA sequence analysis.)

8. Illustrate how everyday computational objects work. (Medicine and other practical areas of science increasingly will be organized and implemented with computational methods. Knowing how some of the programs and databases work is empowering.)

Bibliography

Agee, Jon. Elvis Lives and Other Anagrams. New York: Farrar, Straus and Giroux, 2000.

Aho, A. V. and Ullman, J. D. *Foundations of Computer Science, C Edition*. New York: W. H. Freeman and Company, 1995.

Alpers, Svetlana. *The Art of Describing: Dutch Art in the Seventeenth Century*. Chicago: University of Chicago Press, 1984.

Altschul, S. F., Gish, W., Miller, W., Myers, E. W., and Lipman, D. J. "Basic local alignment search tool." *Journal of Molecular Biology* 215 (1990): 403–410.

Augarde, T. *The Oxford Guide to Word Games*, 2nd ed. New York: Oxford University Press, 2003.

Babbage, Charles. Quoted in "Recreations of a Philosopher." *Harper's* 30 (1864):34–39.

Baum, D. A., Smith, S. D., and Donovan, S. S. "The tree-thinking challenge." *Science* 310 (2005): 979–980.

Birk, J. B., ed. *Rutherford at Manchester*. London: Heywood, 1962.

Byatt, A. S. "Introduction." In *Indexers and Indexes in Fact and Fiction*, ed. Hazel Bell. Toronto: University of Toronto Press, 2001.

Casti, J. *Complexification*. New York: Harper Perennial, 1995.

Chargaff, E. "Preface to a grammar of biology." *Science* 171 (1971): 637–642.

Chargaff, E. *Essays on Nucleic Acids*. New York: Elsevier, 1963.

Conway, Damian. *Perl Best Practices*. Sebastopol, CA: O'Reilly, 2005.

Cross, D. *Data Munging with Perl*. Greenwich, CT: Manning Publications, 2001.

Davis, E. "Divine intervention: an interview with perl creator Larry Wall." *Feed* (February 10, 1999).

Dawkins, R. *The Blind Watchmaker: Why the Evidence of Evolution Reveals a Universe Without Design*. New York: Norton, 1996.

Doyle, A. C. *The Sign of Four*. 1890; New York: Penguin Classics, 2001.

Durbin, R., Eddy, S. R., Krogh, A., and Mitchison, G. *Biological Sequence Analysis: Probabilistic Models of Proteins and Nucleic Acids*. Cambridge: Cambridge University Press, 1998.

Dwyer, R. A. *Genomic Perl: From Bioinformatics Basics to Working Code*. Cambridge: Cambridge University Press, 2003.

Dyer, B. D., LeBlanc, M. D., Benz, S., Cahalan, P., Donorfio, B., Sagui, P., Villa, A., and Williams, G. "A DNA motif lexicon: cataloguing and annotating sequences." *In Silico Biology* 4, 39 (2004): 471–478.

Friedl, J. E. F. *Mastering Regular Expressions*, 2nd ed. Sebastopol, CA: O'Reilly, 2002.

Green, D. *The Serendipity Machine: A Voyage of Discovery Through the Unexpected World of Computers*. New York: Allen and Unwin, 2005.

Hammock, E. and Young, L. J. "Microsatellite instability generates diversity in brain and sociobehavioral traits." *Science* 308 (2005): 1630–1634.

Kaplan, R. and Kaplan, E. *The Nothing That Is: A Natural History of Zero*. New York: Oxford University Press, 1999.

Krane, D. E. and Raymer, M. L. *Fundamental Concepts of Bioinformatics*. San Francisco: Benjamin Cummins, 2003.

Lesk, A. M. *Introduction to Bioinformatics*, 2nd ed. New York: Oxford University Press, 2005.

Levy, D. *Scrolling Forward: Making Sense of Documents in the Digital Age*. New York: Arcade, 2003.

Margolis, J. and Fisher, A. *Unlocking the Clubhouse: Women in Computing*. Cambridge, MA: MIT Press, 2002.

Moorhouse, M. and Barry, P. *Bioinformatics, Biocomputing, and Perl*. Chichester, U.K.: Wiley, 2004.

Palaima, T. G. and Trombley, S. "Archives revive interest in forgotten life." *Austin American-Statesman*, 10/27/2003, p. A9.

Paydarfer, D. and Schwartz, W. "An algorithm for discovery." *Science* 292 (2001): 13.

Pickover, C., ed. *The Pattern Book: Fractals, Art, and Nature*. Singapore: World Scientific Publishing, 1995.

Seife, C. Zero: *The Biography of a Dangerous Idea*. New York: Penguin, 2000.

Srinivasan, S. *Advanced Perl Programming*. Sebastopol, CA: O'Reilly, 1997.

Stein, L. "How Perl Saved the Human Genome Project." In *Games, Diversions, and Perl Culture*, ed. Jon Orwant. New York: O'Reilly, 2004.

Stevenson, R. L. *Virginibus puerisque, and other papers*. New York: Scribner's, 1902.

Tsonis, P. *Anatomy of Gene Regulation: A Three-Dimensional Structural Analysis*. Cambridge: Cambridge University Press, 2003.

Wall, L., Christiansen, T., and Orwant, J. *Programming Perl*, 3rd ed. Sebastopol, CA: O'Reilly, 2000.

Wheatley, H. B. *What Is an Index?* London: Index Society Publication, 1878.

Winchester, S. *The Meaning of Everything*. New York: Oxford University Press, 2003.

Index